Springer Theses

Recognizing Outstanding Ph.D. Research

Aims and Scope

The series "Springer Theses" brings together a selection of the very best Ph.D. theses from around the world and across the physical sciences. Nominated and endorsed by two recognized specialists, each published volume has been selected for its scientific excellence and the high impact of its contents for the pertinent field of research. For greater accessibility to non-specialists, the published versions include an extended introduction, as well as a foreword by the student's supervisor explaining the special relevance of the work for the field. As a whole, the series will provide a valuable resource both for newcomers to the research fields described, and for other scientists seeking detailed background information on special questions. Finally, it provides an accredited documentation of the valuable contributions made by today's younger generation of scientists.

Theses may be nominated for publication in this series by heads of department at internationally leading universities or institutes and should fulfill all of the following criteria

- They must be written in good English.
- The topic should fall within the confines of Chemistry, Physics, Earth Sciences, Engineering and related interdisciplinary fields such as Materials, Nanoscience, Chemical Engineering, Complex Systems and Biophysics.
- The work reported in the thesis must represent a significant scientific advance.
- If the thesis includes previously published material, permission to reproduce this must be gained from the respective copyright holder (a maximum 30% of the thesis should be a verbatim reproduction from the author's previous publications).
- They must have been examined and passed during the 12 months prior to nomination.
- Each thesis should include a foreword by the supervisor outlining the significance of its content.
- The theses should have a clearly defined structure including an introduction accessible to new PhD students and scientists not expert in the relevant field.

Indexed by zbMATH.

More information about this series at https://link.springer.com/bookseries/8790

Fabian Guignard

On Spatio-Temporal Data Modelling and Uncertainty Quantification Using Machine Learning and Information Theory

Doctoral Thesis accepted by
University of Lausanne, Switzerland

Springer

Author
Dr. Fabian Guignard
Institute of Earth Surface Dynamics
Faculty of Geosciences and Environment
University of Lausanne
Lausanne, Switzerland

Supervisor
Prof. Mikhail Kanevski
Institute of Earth Surface Dynamics
Faculty of Geosciences and Environment
University of Lausanne
Lausanne, Switzerland

ISSN 2190-5053 ISSN 2190-5061 (electronic)
Springer Theses
ISBN 978-3-030-95230-3 ISBN 978-3-030-95231-0 (eBook)
https://doi.org/10.1007/978-3-030-95231-0

This Springer imprint is published by the registered company Springer Nature Switzerland AG
The registered company address is: Gewerbestrasse 11, 6330 Cham, Switzerland

N'a de convictions que celui qui n'a rien approfondi.

Emil Cioran

To my parents, Ida and Aimé

To Sandie and our children, Samuel and Elliot

Supervisor's Foreword

The amount of data and information available in geo- and environmental sciences is steadily increasing due to the intensive development of monitoring networks, field campaigns, Earth observation by remote sensing devices (satellites, drones, lidars), and modelling results. Therefore, the efforts in technological advances in data acquisition, storage, and access play an important role. Besides, efficient treatment and use of these data is crucial in scientific developments and intelligent decision-making processes in significant fields such as climate change, environmental risks and natural hazards, and renewable energy assessment. Analysis and modelling of spatio-temporal data are challenging because they are high dimensional and variable at many spatio-temporal scales, multivariate, non-homogeneous, non-linear, and uncertain.

Fabian Guignard was involved in a project of the 75th National Research Program funded by the Swiss National Science Foundation, entitled "Hybrid renewable energy potential for the built environment using big data: forecasting and uncertainty estimation". One of the project's fundamental objectives was to develop data-driven approaches for exploring, analysing, and modelling renewable energy potential in complex mountainous regions. Accordingly, Fabian's thesis resulted in interdisciplinary research combining applied mathematics, spatio-temporal geostatistics, analysis of complex and high-frequency time series using information measures, computer science and machine learning (ML), quantification of data and model uncertainties. An innovative contribution deals with developing and adapting a generic ML-based methodology for spatio-temporal data analysis using Extreme Learning Machine algorithm, for which analytical formulas of the bias and variance are derived. An applied part of the work treats a variety of real environmental data case studies.

This book presents a complete data science workflow: from deriving new theoretical results, via implementing the developed algorithms and methods in software packages, to applying them to concrete environmental applications, and helping extract valuable knowledge from data. I congratulate Fabian on his outstanding work, and I am pleased to announce that he has been awarded the Faculty of Geosciences and Environment Award from the University of Lausanne (UNIL).

Lausanne, Switzerland
December 2021

Prof. Mikhail Kanevski

Acknowledgements

This thesis would not have the same flavour without numerous great scientists and engineers I have met during the past four years, some of whom have become close friends.

First, I would like to thank especially Prof. Mikhail Kanevski, my supervisor. During this thesis, I benefited from his eclectic knowledge of data science. From the beginning, he considered me a researcher and gave me the freedom that I needed.

As a mathematician by training, I am used to living in an abstract world. Fortunately, I met Dr. Federico Amato and Dr. Mohamed Laib, who have brought me back to reality more than once when necessary, and supported me literally until the very last day of this adventure. Both have lived the FiSh and UncELMe adventures with me. I also benefited from the guidance of Dr. Jean Golay, who greatly inspired me from a scientific perspective during my darkest hours. They receive my gratitude here.

I would like to thank the Jury members, Prof. Dionissios Hristopulos, Prof. François Bavaud, and Prof. Antoine Guisan, for agreeing to read and review the manuscript and endorse it for publication. I also thank the thesis advisory committee composed of Prof. David Ginsbourger and Dr. Christian Kaiser for their guidance.

I owe a debt to all my co-authors, with whom I always had fruitful collaborations. My special thanks go to Dr. Luciano Telesca and Dr. Dasaraden Mauree for their insights on environmental physics. I would like to acknowledge Dr. Sylvain Robert for his statistical advice. My thanks also go to the EPFL HyEnergy team, Dr. Alina Walch, Prof. Jean-Louis Scartezzini, Dr. Roberto Castello, and Dr. Nahid Mohajeri.

This thesis would not have been possible without the financial support of the Swiss National Science Foundation (SNSF). This research was conducted within the HyEnergy project n° 167285, which belongs to the 75th National Research Programme (NRP75) of the SNSF. Moreover, the entire work has been done at the Institute of Earth Surface Dynamics of the University of Lausanne, which provides all necessary support. Special thanks to the data providers, MeteoSwiss, Swisstopo, the Solar Energy and Building Physics Laboratory LESO-PB (EPFL), the WSL Institute for Snow and Avalanche Research SLF and the Federal Office for the Environment for allowing me to study various environmental phenomena.

Finally, this research would not have been possible without my family, Nonna, Papou, Papy Marcel, Antoinette, the Thursday and Sunday victuals and the red wine. Samuel and Elliot, it is a wonderful gift to see you grow up. I never know what you will do next minute, pure nugget effect and no machine learning prediction possible for you. Sandie, none of this would have been possible without you. Thank you for your sense of humour and your patience; you have more than you believe.

Lausanne, Switzerland
December 2021

Fabian Guignard

Contents

Abbreviations

ACF	Auto Correlation Function
ANN	Artificial Neural Network
APV	A Priori Variance
BHM	Bayesian Hierarchical Model
BR	Bias-Reduced Estimate
CI	Confidence Intervals
DEM	Digital Elevation Model
DL	Deep Learning
DoG	Difference of Gaussians
DSTM	Dynamic Spatio-Temporal Model
DWT	Discrete Wavelet Transform
EDA	Exploratory Data Analysis
ELM	Extreme Learning Machine
EOF	Empirical Orthogonal Function
EPFL	Ecole Polytechnique Fédérale de Lausanne
FIM	Fisher Information Measure
FNN	Feed-Forward Neural Network
FSC	Fisher-Shannon Complexity
FSIP	Fisher-Shannon Information Plane
GCV	Generalised Cross-Validation
HCCME	Heteroskedastic-Consistent Covariance Matrix Estimator
HyREP	Hybrid Renewable Energy Potential
i.i.d.	Independent and Identically Distributed
IT	Information Theory
KDE	Kernel Density Estimator
LS	Least-Squares
MAE	Mean Absolute Error
ML	Machine Learning
MLR	Multiple Linear Regression
MoTUS	Measurement of Turbulence in a Urban Setup
MRWA	Multiresolution Wavelet Analysis

MSE	Mean Square Error
NHe	Naive Heteroskedastic Estimate
NHo	Naive Homoskedastic Estimate
NRP75	National Research Programme 75
PC	Principal Component
PCA	Principal Component Analysis
PDF	Probability Density Function
PSD	Power Spectral Density
Q-Q	Quantile-Quantile
RE	Relative Mean Square Error
RHS	Right-Hand Side
RMSE	Root Mean Square Error
RSS	Residual Sum of Squares
SEP	Shannon Entropy Power
SNSF	Swiss National Science Foundation
SVD	Singular Value Decomposition
UNIL	UNIversité de Lausanne
UQ	Uncertainty Quantification

Chapter 1
Introduction

Spatio-temporal data are everywhere. Scientific observations are spatio-temporal by nature in many problems coming from various fields such as geo-environmental sciences, electricity demand management, geomarketing, neurosciences, public-health, or archaeology. This thesis is focused on methods that visualise, explore, unveil patterns, and predict such spatio-temporal data. Applications of the proposed methodologies are specifically concentrated on data from environmental sciences, with an emphasis on wind speed modelling in complex mountainous terrain and the resulting renewable energy assessment.

In this chapter, Sect. 1.1 briefly introduces the main topics of the thesis. Section 1.2 deepens the general context. The objectives are specified in Sect. 1.3. Section 1.4 provides thesis contributions that address these objectives. Section 1.5 gives the outline of the thesis and lists all the published sources on which the manuscript is based.

1.1 Sketch of the Thesis

Spatio-temporal data generated by environmental phenomena can be very complex. Depending on the temporal frequency of the study, the data can show temporal non-stationarity, seasonalities, high variability at multiple scales and complicated behaviour. This behaviour can change in time, but also across space. Although these data are known for temporal and spatial dependencies, they can also hide spatio-temporal interactions in their structures. Moreover, complex spatio-temporal phenomena are often generated by non-linear processes.

First, to gain understanding from such complicated spatio-temporal data and uncover real data patterns, effective and efficient exploratory tools are necessary. Information Theory (IT) offers a convenient framework and can provide numerous insights into the data. The thesis investigates the joint use of Fisher Information

F. Guignard, *On Spatio-Temporal Data Modelling and Uncertainty Quantification Using Machine Learning and Information Theory* , Springer Theses,
https://doi.org/10.1007/978-3-030-95231-0_1

Measure (FIM) and Shannon Entropy Power (SEP), two IT-related quantities, for assessing the complexity of distributional properties of temporal, spatial and spatio-temporal data sets.

Second, being universal modelling tools, Machine Learning (ML) algorithms are well suited to model non-linear environmental phenomena. Interpolation in the geographical space is obtained through prediction in a higher-dimensional feature space. However, with complex spatio-temporal data, high-frequency behaviour could be a problem. This is overcome by describing the data in a convenient basis. The thesis elaborates an effective ML-based framework for modelling spatio-temporal fields measured on monitoring networks that can take a relatively high number of spatial features and data into account.

Third, while ML can achieve excellent performance in prediction, major questions arise in terms of Uncertainty Quantification (UQ) in most cases. The thesis investigates the UQ feasibility rigorously for Extreme Learning Machine (ELM)—a particular type of neural network—under the frequentist paradigm.

Finally, the thesis provides a wide range of application examples of the proposed methodology, with a particular focus on the spatio-temporal field of wind speed in Switzerland, aiming to produce a renewable energy potential assessment at a national scale. The proposed spatio-temporal data modelling framework is compatible with the newly introduced UQ method, which measures the prediction belief and improves decision-making.

1.2 Spatio-Temporal Data Modelling and Analysis

According to [1], statistical modelling of a spatio-temporal data set pursues three principal goals. The first one is to predict a value at a new spatio-temporal point within the spatio-temporal "cube" defined by the spatial and temporal bounds of the given measurements. This task is also referred to as interpolation problem. Accompanying this prediction, it is often requested to provide its uncertainty. The second one is about scientific inference on the relationship between predictors and the predicted variable, often through parameter inference. Finally, the third one is to forecast a future value at some spatial location, i.e. outside the spatio-temporal "cube". Here again, the UQ of the forecasting can be of great interest.

The thesis deals mainly with the first goal, although some aspects of the two others could be mentioned. To make the interpolation task feasible, a hypothesis on spatio-temporal continuity is required [1, 2]. That is, nearby measurements tend to be similar more than distant ones, which is known as Tobler's law in geography [3].

1.2.1 Statistical Interpolation for Spatio-Temporal Data

Geostatistics is a popular and well-established field [4–6]. The family of kriging models is the workhorse for geostatistical spatial prediction. It estimates the interpolated value as an optimal linear weighted combination of neighbours, in the sense that the weights are chosen such as the prediction is unbiased and has a minimum variance. These optimal weights are obtained by considering and modelling the spatial covariance function, which describes the covariance structure present in the random field that supposed to generate the data. This task is usually done through the variogram, which measures spatial statistical linear dependencies of the phenomena under study. Studying the different properties of the covariance function, such as isotropy and strict positive definiteness, is crucial for successful modelling with kriging. Geostatistics has proved its effectiveness in many situations, even with a small amount of data. However, in most cases, complex non-linear phenomena are considered in a high-dimensional heterogeneous feature spaces [7] and not in geographical space only [8]. In this case, ML is sometimes considered preferable [2].

Spatio-temporal geostatistics is the natural generalisation of geostatistics that make come into play the temporal dimension. Most of the involved concepts are analogous, such as kriging, covariance function and variogram, but the covariance structure describing the statistical linear dependencies is spatio-temporal [9–11]. Additional properties of the covariance function become of a capital interest, such as symmetry and separability [12]. The latter is of particular interest for mathematical convenience. When time is considered, the number of observations multiplies. As the mathematical mechanism requires the inversion of a square matrix of a size determined by the total number of observations, kriging can be infeasible. However, several procedures exist to overcome this [13]. One possibility is to assume separability, that is, the spatio-temporal covariance function can be characterized by the product of purely temporal and purely spatial covariance functions. This allows the emergence of simplifications through the Kronecker product, which reduces the computational burden. Moreover, the entire spatio-temporal linear dependencies can be described from its spatial and its temporal projections, which is also very convenient for modelling purposes [9]. However, depending on the phenomena under study, this assumption is not always tenable [12, 14], although a nearest Kronecker product approximation of non-separable covariance has been developed [15]. In addition to these computational concerns, dealing with stationarity and complex temporal behaviour can be very challenging. Moreover, covariance-based models are not sufficient to characterise spatio-temporal dependencies within non-linear phenomena [1].

When complex situations arise, Bayesian Hierarchical Models (BHMs) are sometimes recommended for predicting and forecasting spatio-temporal data. This is especially true when strong mechanistic knowledge about the process is known. The unified probabilistic framework of BHM—which allows UQ of the data, the model and its parameters—is sometimes presented as a potential advantage compared to deep learning models [1]. Dynamic Spatio-Temporal Models (DSTMs) can be built following the BHM framework [10]. DSTMs can have non-linear extensions, although this kind of model requires strong expertise in the domain.

1.2.2 Machine Learning for Spatial Data

Although ML is not designed for, it can solve environmental problems [2, 16]. Adaptations of ML algorithms for spatial interpolation tasks were proposed, such as artificial neural networks [17–20] and kernel methods [21–23]. Most ML algorithms are non-linear models and universal approximators, which loosely speaking means that they can approximate any non-linear function with desired precision. However, their main advantage is probably that they can handle high-dimensional input space in which they can efficiently find acceptable solutions [24]. In the spatial interpolation context, the input space is constructed by adding relevant spatially referenced information to the geographical space [2]. As an example, wind speed monthly averages are highly influenced by the local topographic shape [7]. In addition to location, ML algorithms allow considering a high number of extra features describing the terrain. Some processes are dominated mainly by external variables, making the use of classical methods difficult. When it happens, the high-dimensionality aspect of the input space of ML methods becomes indispensable [2].

It is usually stated that the choice of the approach is a trade-off between flexibility and interpretability [25]. If prediction accuracy is the main goal pursued, a certain flexibility of the method is often desired. On the contrary, if the inference is the main interest of the study at hand, a less flexible but more interpretable approach would be preferred. Generally speaking, ML algorithms belong to the highly flexible but not very interpretable approaches.

Some links exist between geostatistics and ML. On the one hand, kernel-based methods and geostatistics are related, as kernels are covariance functions [26]. Notably, kriging can be seen as a particular case of Gaussian processes using geographical space—and potentially time—as an input space of dimension 2 or 3 [27]. On the other hand, it can be shown that a single layer neural network converges to a Gaussian process as the number of neurons goes to infinity [27–29]. It suggests that neural networks could be closer to kriging models than it might seem.

1.2.3 Uncertainty Quantification

Given any prediction, it is always convenient—if not necessary—to measure its belief in it. In environmental risk assessments and decision-making processes, UQ is extremely important and sometimes even more than the prediction itself [2]. UQ relies on the validity of the model, and it cannot claim if the model is wrong or not [30]. Nevertheless, it can provide crucial conclusions on the data and their impact on the model and its parameters. Common uncertainty summary measures are variance-based, such as the covariance matrix, its trace, its determinant, or in the univariate case, the variance itself [30, 31].

Here we will concern about the aleatoric conception of uncertainty, which is the one adopted by, e.g. the frequentist paradigm and typically involved random outcomes of a repeated experiment, differently from the epistemic conception of uncertainty, which is the one adopted by, e.g. the Bayesian paradigm and typically involved missing information or expertise. Epistemic uncertainty is often considered as a subjective measure of belief [32, 33].

In general, UQ is far much better rooted in the statistical culture than in ML. While their prediction accuracy can be surprisingly good, ML methods are often quite limited in UQ, even to get a simple variance estimate. UQ method for ML mainly relies on bootstrap [25], which is an effective proven tool [34]. However, in the context of spatio-temporal dependencies, bootstrapping can be very challenging [1], although procedures for purely temporal or spatial data are well-documented [35].

Among spectral decompositions of random variables, Karhunen–Loéve expansion is of particular interest [36]. It can decompose a spatio-temporal random field in a bi-orthogonal fashion, i.e. the random field is expressed as a linear combination of deterministic orthogonal basis with stochastic centred coefficients that are also orthogonal in the probability space, hence uncorrelated. This property is appreciated in the context of UQ [30]. Moreover, such bases are optimal, in the sense that truncated expansion minimises the Mean Square Error (MSE) [36].

1.2.4 Information Theory as an Advanced Exploratory Tool

IT is based on applied mathematics, computer science, electrical engineering, and physics [37]. It finds many applications to diverse areas, including statistics, ML, dynamical systems and signal processing. A fundamental quantity in IT is entropy, which can also be seen as a scalar summary measure for UQ [30]. Actually, for a Gaussian random vector, its entropy and its covariance matrix determinant are equivalent—up to a strictly monotone transform [31]. Another quantity appearing in IT is FIM, although it stems in mathematical statistics [37, 38]. Last decade, increasing attention is paid towards the Fisher-Shannon analysis, a method using FIM and a non-linear transformation of entropy—SEP—to characterise the complexity and non-stationarity of non-linear time series. A simple combination of these two quantities was proposed as a statistical complexity measure, namely, the Fisher-Shannon Complexity (FSC) [39, 40].

1.3 Objectives

Although the objectives and contributions of this thesis are mainly methodological, their motivations stem from a real-world problem, namely an aeolian energy assessment. As for many other environmental ground measurements, available wind data—that will be called the MeteoSwiss data—come from monitoring networks covering Switzerland territory with a fixed temporal sampling frequency, which produces a specific type of data structure: they are discrete and regular in time while continuous but irregularly sampled in space [1]. This structure directly impacts the type of method that can be used to reach the interpolation goal. These data are completed by a very high-frequency wind data set—that will be called the MoTUS data—which will be exclusively devoted to the Exploratory Data Analysis (EDA) to gain understanding of the wind phenomena behaviour.

Data exploration is crucial at every step of the modelling, from raw data to residual analysis. The extensive growth of complicated non-linear data requires developing reliable techniques to extract knowledge from them and understand phenomena under study. Quantifying the complexity and non-stationarity of complicated non-linear time series is challenging, and advanced complexity-based exploratory tools are required to understand and visualise such data. The first objective is the following.

> **Objective 1**
> Investigate and develop the use of the Fisher-Shannon analysis, in particular the FSC, as an advanced EDA tool to assess the complexity of distributional properties of temporal, spatial and spatio-temporal data sets.

Given the nature and the complex behaviour of the spatio-temporal data, it would be convenient to get ML into the spatio-temporal interpolation problem. The model would benefit from a higher dimensional input space, which enables the use of many external heterogeneous geo-features. Therefore, developing new spatio-temporal models is extremely important both from scientific and applied points of view.

> **Objective 2**
> Propose and elaborate a methodological framework adaptable to different ML algorithms for modelling complex spatio-temporal data irregularly sampled in space with a fixed temporal high-frequency.

Once the methodological framework is established, it should be combined with a particular ML method for which an effective and efficient UQ technique can be developed.

> **Objective 3**
> For a well-chosen ML algorithm, develop a reliable UQ method established on a solid statistical basis.

The final aim of any data analysis methodology is rooted in the real world, and the developments proposed to address the three aforementioned objectives will be applied to a case study, which is presented here.

The HyEnergy project aims to estimate the Hybrid Renewable Energy Potential (HyREP) to supply Swiss buildings and quantify the uncertainty of the estimation. HyREP systems combine solar, wind and shallow geothermal energy, reducing energy storage capacity and operating costs. The main reason for hybridising these three renewable energy resources is that they are supposed to be complementary—to a certain point. Wind energy increases during overcast and precipitation, compensating for the decreasing of solar productivity. The yearly cycle is essential for these energies, but the daily cycle is also crucial for solar energy. Differently, geothermal energy production does not depend on the time of the year or the day, nor weather changes. It is assumed to serve as a backup during calm and cold winter, when solar and wind energy production is low, while during summer, it is used to cool down houses and photovoltaics.

To provide such assessment is important to support decision-making processes and policies for buildings. Interpolation of environmental variables is a crucial step in this estimation. In particular, the thesis author's task was *in fine* to help to provide an estimate and its uncertainty for wind speed, for any location in Switzerland and every hour during ten consecutive years, in order to proceed to energy potential computations. The temporal frequency is motivated by the practical considerations stated above.

Wind phenomena is a non-linear process that is turbulent by nature. According to the temporal frequency of study, it produces non-stationary wind speed time series with complex high-frequency behaviour, high variability and seasonalities. In a complex topographic region such as Switzerland, the wind is also strongly influenced by terrain [41], and wind speed may show spatial trends depending on the climatic regions. Moreover, the wind speed spatio-temporal field involves temporal, spatial, and non-separable spatio-temporal dependencies, making wind speed analysis and modelling challenging tasks, especially at such frequency.

Objective 4

Apply the developments proposed to address the three aforementioned objectives, which in terms of the HyEnergy project requirements correspond to:

- Improve understanding of the wind phenomenon in free spaces and urban areas based on the MoTUS and MeteoSwiss data.
- Modelling wind speed on a regular grid over Switzerland for every hour during ten consecutive years and quantify the uncertainty of such modelling.
- Estimate wind energy potential and its uncertainty.

1.4 Contributions

The use of Fisher-Shannon analysis as an advanced EDA tool is investigated. Particular attention is paid to the FSC measure. The state-of-the-art coming from various fields are collected, analytical development and new interpretations are provided, and software for non-parametric estimations are developed. The high versatility and usefulness of the FSC, and more generally the Fisher-Shannon method, are demonstrated through numerous case studies based on simulated and real-world data. The main contributions can be summarised as follows:

Contribution to Fisher-Shannon analysis and FSC as an EDA tool

- Identification of FSC as a sensitivity measure of the SEP, as a scale-independent non-Gaussianity measure of data and potentially as a multi-modality measure.
- Derivation of some new analytical results on FIM and FSC.
- Software development in Python and R.

In order to interpolate the spatio-temporal data, it is proposed to decompose them into a linear temporal basis with spatial stochastic coefficients. The spatial coefficients are interpolated in space with ML algorithms and then recomposed, predicting the spatio-temporal field at a new location in space at the same temporal frequency. It enables the modelling of non-stationary, complex seasonal and high frequency, space-time processes. Note that this idea is not new [1], but was never proposed with ML, up to the author's knowledge.

Consequently, the spatio-temporal model can consider high dimensional input space of spatial covariates and manage large data sets, with high frequency in time but with few spatial points—such as data produced by a monitoring network. An example of temporal basis is provided by the bi-orthogonal Karhunen–Loève decomposition, which corresponds to Empirical Orthogonal Functions (EOFs) in the discrete case, also known as Principal Component Analysis (PCA). EOFs basis is entirely data-driven and specifically chosen for its property of MSE minimisation of truncated expansion, which will enable dimensionality reduction of the data in an optimal way.

Contribution to non-linear spatio-temporal interpolation
Design of an ML framework to interpolate spatio-temporal data considering the sampling, irregular in space and regular in time, and the relatively complex non-stationary temporal behaviour.

At this stage, it turns out that it is possible to derive analytical properties and variance estimates for a particular type of Artificial Neural Network (ANN) with random input weights—ELM. Basically, this algorithm is a multiple linear regression performed in a random feature space. However, most UQ methods proposed in the literature for ELM make strong assumptions on the data, ignore the randomness of input weights, or neglect the bias contribution in confidence interval estimations.

Here, novel estimations that overcome these constraints and improve the understanding of ELM variability are presented. Analytical derivations are provided under general assumptions, supporting the identification and interpretation of the contribution of different variability sources. Under both homoskedasticity and heteroskedasticity, several variance estimates are proposed, implemented, investigated, and numerically tested, showing their effectiveness in replicating the expected variance behaviours. Finally, the feasibility of confidence intervals estimation is discussed. The Tikhonov regularised version of ELM is also included in all these investigations.

These contributions have a general scope and are not specific to the spatio-temporal interpolation problem. Variance estimates can be applied to any non-linear regression problem solved by ELM.

Contribution to UQ for ELM

- Development of analytical results, helping clarify the impact of the different variability source on ELM and increasing its understanding, including Tikhonov regularised version.
- Development of several homoskedastic and heteroskedastic variance estimates, investigation of their properties and extensive testing with numerical simulations.
- Investigations on the feasibility of confidence interval estimations.
- Software development in Python compatible with scikit-learn.

These three methodological developments are applied to the wind speed data sets mentioned above, contributing to renewable energy assessment while providing a comprehensive example of the overall methodology.

The data sets are explored with classical tools such as visualisation and variography. Fisher-Shannon tools are extensively used, particularly on the high-frequency MoTUS dataset, yielding numerous insights on wind speed. Globally, EDA shows the extent of the difficulty to model wind speed data in a mountainous region at such temporal frequency, which appears to be a real challenge.

Based on the MeteoSwiss data, hourly wind speed is estimated on a spatial grid of $250 \times 250\,\mathrm{m}^2$ covering Switzerland. The data are decomposed in EOFs, and the spatial coefficients are modelled with ELM, which provides UQ for the spatio-temporal prediction. EDA tools are then reused on residuals to provide modelling assessment. Finally, the results are converted into aeolian energy potential and uncertainty is propagated.

Contribution to renewable energy assessment in Switzerland

- Comprehensive EDA of the available wind speed data sets, providing various insights which contribute to a better understanding of the wind phenomenon in free and urban spaces.
- Estimation of hourly wind speed in Switzerland.
- UQ of the proposed prediction.
- Conversion of the wind speed interpolation results in aeolian energy potential.

1.5 Thesis Organisation

The book encompasses eight chapters that follow the flow of a classical data analysis process. The proposed methodologies are presented along with applications to data as examples. Visualisation accompanies the exploration of the data and the analysis results at each step.

Chapter 2 presents the study area and the wind speed data sets, including data gathering, wrangling, cleaning and missing values imputation. The data are also explored using basic EDA tools. Chapter 3 continues exploration with more advanced tools from spatio-temporal statistics and time series. Methodological and practical aspects of the Fisher-Shannon analysis are introduced in Chap. 4, including numerous applications.

A good picture of the data is provided at this stage, and the reader should be convinced of the difficulty in modelling them. The next three chapters are devoted to the interpolation task. Chapter 5 introduces the spatio-temporal interpolation framework, which is tested on simulated and real-world data. Original developments about UQ with ELM and extensive numerical simulations are presented in Chap. 6. These are combined with the interpolation framework and applied to wind speed in Switzerland in Chap. 7. Particular attention is paid to the analysis of the residuals, which confirms that the proposed framework does reasonable work. This chapter also presents how the results can help in estimating the energy potential. Finally, Chap. 8 concludes the thesis and proposes future perspectives.

Abbreviations are defined at their first appearance. A list of abbreviations can be found at the beginning of the book. Words written in *italic* emphasise definitions of concepts, data sets, or quantities that may be new to the reader.

The thesis is based on several publications—listed below—for which the thesis author provided a substantial scientific contribution, if not the main. This book integrates, partially or totally, their methodological developments, analysis, results and written contents. In particular, some passages have been quoted *verbatim*, while other passages have been adapted or paraphrased from them. Several co-authors—from the Institute of Earth Surface Dynamics (UNIL, Switzerland), the Solar Energy and Building Physics Laboratory (EPFL, Switzerland), the Institute of Methodologies for Environmental Analysis (CNR, Italy), the Luxembourg Institute for Science and Technology (LIST, Luxembourg), the California Institute of Technology (Caltech, USA), the Institute for Environmental Design and Engineering (UCL, UK), the Institute for Snow and Avalanche Research (WSL, Switzerland) and the Regional Agency for the Protection of the Environment of Basilicata (ARPAB, Italy)—are involved and have provided contributions.

For transparency, the thesis author contributions are specified. The peer-reviewed articles appearing in the thesis are the following:

[42] *Uncertainty Quantification in Extreme Learning Machine: Analytical Developments, Variance Estimates and Confidence Intervals* by F. Guignard, F. Amato, and M. Kanevski. Published in Neurocomputing, vol. 456, pp. 436–449, 2021.
The thesis author conceived the main conceptual ideas, conduct investigations, developed the theoretical formalism and the methodology, designed the experiments, performed the calculations, interpreted the computational results, wrote the original draft, and developed the Python software. He also contributes to discussed the results, provided critical feedback, commented, reviewed, edited and corrected the final version of the paper.

[43] *Spatio-temporal Estimation of Wind Speed and Wind Power using Machine Learning: Predictions, Uncertainty and Policy Indications* by F. Guignard, F. Amato, A. Walch, R. Castello, N. Mohajeri, J.-L. Scartezzini, M. Kanevski. Under review.
The thesis author contributes to conceived the main conceptual ideas, preprocessed the data, conduct investigations, developed the theoretical formalism and the methodology, designed the experiments, conduct the residuals analysis, performed the calculations, interpreted the computational results, wrote the original draft, discussed the results and provided critical feedback.

[44] *A Novel Framework for Spatio-Temporal Prediction of Environmental Data using Deep Learning* by F. Amato, F. Guignard, S. Robert, M. Kanevski. Published in Scientific Reports, vol. 10, no. 1, 22243, 2020.
The thesis author contributes to conceived the main conceptual ideas, preprocessed the data, conduct the residuals analysis, discussed the results, provided critical feedback, wrote the original draft, commented, reviewed and edited the original manuscript.

[45] *Advanced Analysis of Temporal Data using Fisher-Shannon Information: Theoretical Development and Application in Geosciences* by F. Guignard, M. Laib, F. Amato, and M. Kanevski. Published in Frontiers in Earth Science, vol. 8, 2020.
The thesis author conceived the main conceptual ideas, conduct investigations, developed the theoretical formalism, performed the calculations, interpreted the computational results, and wrote the original draft. He also contributes to discussed the results, provided critical feedback, commented, reviewed and edited the original manuscript, wrote the final version of the paper and developed Python and R packages.

[46] *Fisher-Shannon Complexity Analysis of High-Frequency Urban Wind Speed Time Series* by F. Guignard, D. Mauree, M. Lovallo, M. Kanevski, and L. Telesca. Published in Entropy, vol. 21, no. 1, p. 47, 2019.
The thesis author preprocessed the data and contributes to conduct investigations, performed the calculations, interpreted the computational results, wrote the original draft, discussed the results and provided critical feedback, commented, reviewed, edited and corrected the final version of the paper.
[47] *Wavelet Variance Scale-Dependence as a Dynamics Discriminating Tool in High-Frequency Urban Wind Speed Time Series* by F. Guignard, D. Mauree, M. Kanevski, and L. Telesca. Published in Physica A: Statistical Mechanics and its Applications, vol. 525, pp. 771–777, 2019.
The thesis author preprocessed the data and performed the calculations. He also contributes to conduct investigations, interpret and discuss the results, provided critical feedback, wrote the original draft, commented, reviewed, edited and corrected the final version of the paper.

Numerous research results of the thesis were also presented at conferences—not listed here. Peer-reviewed conference papers are the following:

[48] *Model Variance for Extreme Learning Machine* by F. Guignard, M. Laib and M. Kanevski. Published in Proceedings of the 28th European Symposium on Artificial Neural Networks, Computational Intelligence and Machine Learning (ESANN), 2020.
[49] *Spatio-Temporal Evolution of Global Surface Temperature Distributions* by F. Amato, F. Guignard, V. Humphrey, M. Kanevski. Published in Proceedings of the 10th International Conference on Climate Informatics, Association for Computing Machinery (ACM), pp. 37–43, 2020.

The articles [42, 44–46, 49] listed above are open access, published and distributed under the terms of the Creative Commons Attribution 4.0 International Licence (CC BY), which is available at http://creativecommons.org/licenses/by/4.0/.

Python and R libraries accompany the thesis:

- *UncELMe*—Uncertainty quantification of Extreme Learning Machine ensemble—is a Python package proposed on PyPI and GitHub (https://github.com/fguignard/UncELMe). It allows interested users to compute all variance estimates for ELM ensembles discussed in the book. It is built within the scikit-learn estimator framework, enabling the use of all convenient functionalities of scikit-learn [50]. Noise estimations are also returned to enable the construction of prediction intervals.
- *FiShPy*—Fisher-Shannon with Python—is a Python package available on PyPI and GitHub (https://github.com/fishinfo/FiShPy). It allows computing a non-parametric estimation of the SEP, FIM and FSC.
- *FiSh*—Fisher-Shannon—is an R package available on CRAN and GitHub (https://github.com/fishinfo/FiSh) which allows the same estimations. It was done in collaboration with Dr. Mohamed Laib.

For the completeness and coherence of the thesis, some interpretations of the results were provided by the experts, co-authors of the joint papers. Specifically, the comments on atmospheric physics of Dr. Dasaraden Mauree and Dr. Luciano Telesca have been reported to Sects. 2.3.2, 3.3.3 and 4.5.4, highlighting the physical interpretation of the statistical findings. Also, the spatio-temporal interpolation framework is put into practice by combining it with a Deep Learning (DL) model in Sect. 5.2.2, from an original idea by Dr. Federico Amato, who performed the calculations.

References

1. Wikle C, Zammit-Mangion A, Cressie N (2019) Spatio-temporal Statistics with R. Chapman & Hall/CRC the R series. CRC Press, Taylor & Francis Group, Boca Raton
2. Kanevski M, Pozdnoukhov A, Timonin V (2009) Machine learning for spatial environmental data. EPFL Press, Lausanne
3. Tobler WR (1970) A computer movie simulating urban growth in the detroit region. Econ Geogr 46(sup1):234–240
4. Chiles J-P, Delfiner P (2009) Geostatistics: modeling spatial uncertainty, vol 497. Wiley, New York
5. Cressie N (2015) Statistics for spatial data. Wiley, New York
6. Kanevski M, Maignan M (2004) Analysis and modelling of spatial environmental data, vol 6501. EPFL Press, Lausanne
7. Robert S, Foresti L, Kanevski M (2013) Spatial prediction of monthly wind speeds in complex terrain with adaptive general regression neural networks. Int J Climatol 33(7):1793–1804
8. Micheletti N, Foresti L, Robert S et al (2014) Machine learning feature selection methods for landslide susceptibility mapping. Math Geosci 46(1):33–57
9. Montero J-M, Fernández-Avilés G, Mateu J (2015) Spatial and spatio-temporal geostatistical modeling and kriging, vol 998. Wiley, New York
10. Cressie N, Wikle C (2011) Statistics for spatio-temporal data. Wiley, New York
11. Sherman M (2011) Spatial statistics and spatio-temporal data: covariance functions and directional properties. Wiley, New York
12. De Iaco S, Posa D, Cappello C, Maggio S (2019) Isotropy, symmetry, separability and strict positive definiteness for covariance functions: a critical review. Spatial Stat 29:89–108
13. Sun Y, Li B, Genton MG (2012) Geostatistics for large datasets. In: Advances and challenges in space-time modelling of natural events. Springer, Berlin, pp 55–77
14. Gneiting T, Genton MG, Guttorp P (2006) Geostatistical space-time models, stationarity, separability, and full symmetry. In: Statistical methods for spatio- temporal systems. Chapman & Hall, Boca Raton, pp 151–175
15. Genton MG (2007) Separable approximations of space-time covariance matrices. Environ: off J Int Environ Soc 18(7):681–695
16. Hsieh WW (2009) Machine learning methods in the environmental sciences: neural networks and kernels. Cambridge University Press, Cambridge
17. Kanevski M, Arutyunyan R, Bolshov L, Demyanov V, Maignan M (1996) Artificial neural networks and spatial estimation of chernobyl fallout. Geoinformatics 7(1–2):5–11
18. Demyanov V, Kanevsky M, Chernov S, Savelieva E, Timonin V (1998) Neural network residual kriging application for climatic data. J Geogr Inf Decis Anal 2(2):215–232
19. Leuenberger M, Kanevski M (2015) Extreme learning machines for spatial environmental data. Comput & Geosci 85:64–73
20. Li Y, Sun Y, Reich BJ (2020) Deepkriging: spatially dependent deep neural networks for spatial prediction. arXiv:2007.11972

21. Gilardi N, Bengio S (2000) Local machine learning models for spatial data analysis. J Geogr Inf Decis Anal 4:11–28
22. Kanevski M, Canu S (2000) Spatial data mapping with support vector regression. Technical Report, IDIAP
23. Foresti L, Tuia D, Kanevski M, Pozdnoukhov A (2011) Learning wind fields with multiple kernels. Stoch Environ Res Risk Assess 25(1):51–66
24. Hastie T, Tibshirani R, Friedman J (2009) The elements of statistical learning: data mining, inference, and prediction. Springer series in statistics. Springer, Berlin, ISBN: 9780387848846
25. James G, Witten D, Hastie T, Tibshirani R (2013) An introduction to statistical learning, vol 112. Springer, Berlin
26. Genton MG (2001) Classes of kernels for machine learning: a statistics perspective. J Mach Learn Res 2(Dec):299–312
27. Rasmussen CE, Williams CK (2006) Gaussian processes for machine learning. MIT Press, Cambridge
28. Pang G, Yang L, Karniadakis GE (2019) Neural-net-induced gaussian process regression for function approximation and pde solution. J Comput Phys 384:270–288
29. Lee J, Bahri Y, Novak R, Schoenholz SS, Pennington J, Sohl- Dickstein J (2017) Deep neural networks as gaussian processes. arXiv:1711.00165
30. Sullivan TJ (2015) Introduction to uncertainty quantification, vol 63. Springer, Berlin
31. Le ND, Zidek JV (2006) Statistical analysis of environmental space-time processes. Springer Science & Business Media, Berlin
32. Fox CR, Ülkümen G (2011) Distinguishing two dimensions of uncertainty. Perspectives on thinking, judging, and decision making, vol 14
33. Lele SR (2020) How should we quantify uncertainty in statistical inference? Front Ecol Evol 8:35. https://doi.org/10.3389/fevo
34. Efron B, Tibshirani RJ (1994) An introduction to the bootstrap. CRC Press, Boca Raton
35. Lahiri SN (2013) Resampling methods for dependent data. Springer Science & Business Media, Berlin
36. Hristopulos DT (2020) Random fields for spatial data modeling. Springer, Berlin
37. Cover TM, Thomas JA (2006) Elements of information theory (Wiley Series in telecommunications and signal processing). Wiley- Interscience, New York, ISBN: 0471241954
38. Knight K (1999) Mathematical statistics. Chapman and Hall/CRC, Boca Raton
39. Angulo J, Antolín J, Sen K (2008) Fisher-Shannon plane and statistical complexity of atoms. Phys Lett A, 372(5):670–674. ISSN: 0375-9601. https://doi.org/10.1016/j.physleta.2007.07.077
40. Esquivel RO, Angulo JC, Antolín J, Dehesa JS, López-Rosa S, Flores-Gallegos N (2010) Analysis of complexity measures and information planes of selected molecules in position and momentum spaces. Phys Chem Chem Phys 12:7108–7116, 26. https://doi.org/10.1039/B927055H.
41. Whiteman C (2000) Mountain meteorology: fundamentals and applications. Oxford University Press, Oxford
42. Guignard F, Amato F, Kanevski M (2021) Uncertainty quantification in extreme learning machine: analytical developments, variance estimates and confidence intervals. Neurocomputing 456:436–449. ISSN: 0925-2312. https://doi.org/10.1016/j.neucom.2021.04.027
43. Guignard F, Amato F, Walch A et al (2021) Spatio-temporal estimation of wind speed and wind power using machine learning: predictions, uncertainty and policy indications. Under review
44. Amato F, Guignard F, Robert S, Kanevski M (2020) A novel framework for spatio-temporal prediction of environmental data using deep learning. Sci Rep 10(1):22 243. https://doi.org/10.1038/s41598-020-79148-7
45. Guignard F, Laib M, Amato F, Kanevski M (2020) Advanced analysis of temporal data using Fisher-Shannon information: theoretical development and application in geosciences. Front Earth Sci 8. https://doi.org/10.3389/feart.2020.00255
46. Guignard F, Mauree D, Lovallo M, Kanevski M, Telesca L (2019) Fisher-shannon complexity analysis of high-frequency urban wind speed time series. Entropy 21(1):47. https://doi.org/10.3390/e21010047

47. Guignard F, Mauree D, Kanevski M, Telesca L (2019) Wavelet variance scale-dependence as a dynamics discriminating tool in high-frequency urban wind speed time series. Phys A: Stat Mech Appl 525:771–777
48. Guignard F, Laib M, Kanevski M (2020) Model variance for extreme learning machine. In: Proceedings of the 28th European symposium on artificial neural networks, computational intelligence and machine learning, pp 703–708
49. Amato F, Guignard F, Humphrey V, Kanevski M (2020) Spatio-temporal evolution of global surface temperature distributions. In: Proceedings of the 10th international conference on climate informatics, pp 37–43. https://doi.org/10.1145/3429309.3429315
50. Pedregosa F, Varoquaux G, Gramfort A et al (2011) Scikitlearn: machine learning in python. J Mach Learn Res 12(Oct):2825–2830

Chapter 2
Study Area and Data Sets

This chapter presents the study area and the data investigated as examples for the proposed methodologies along the thesis. Data wrangling, cleaning and missing values imputation processes are described. The data are explored with basic EDA tools such as summary statistics, spatial and temporal plots, box plots and kernel density estimates. The chapter also provides some basic meteorological insights. Section 2.1 discusses Switzerland topography and presents a 13-dimensional input space that would attempt to characterise the impact of its local shape on the wind flows. Section 2.2 presents and explores the MeteoSwiss wind speed data set, consisting of ten years of hourly measurements in Switzerland. Section 2.3 briefly describes the MoTUS experiment from which high-frequency wind speed profiles are collected and investigated.

2.1 Switzerland and Its Topography

With an area of 41'285 [km^2], a typical characteristic of Switzerland is its complex terrain. It exhibits high elevation changes, from 198 [m] (Lake Maggiore) to 4634 [m] (Dufourspitze) above sea level. The country is crossed from south-west to north-east by the Alps chain, which forms a natural barrier. Its topography leads to various climatic conditions across the territory [1].

One common geographical partition of Switzerland divides it into three geographical regions, defined as *Jura* mountains (north-west), *Plateau* (central), and *Alps* (south). Each part has different geological and topographical properties. This partition is reported with topographical relief in Fig. 2.1. Most of the Alps chain's highest mountains belong to the Swiss Alps, covering about 60% of the territory. The Plateau occupies about 30% of the country's area and features the main economic

F. Guignard, *On Spatio-Temporal Data Modelling and Uncertainty Quantification Using Machine Learning and Information Theory* , Springer Theses,
https://doi.org/10.1007/978-3-030-95231-0_2

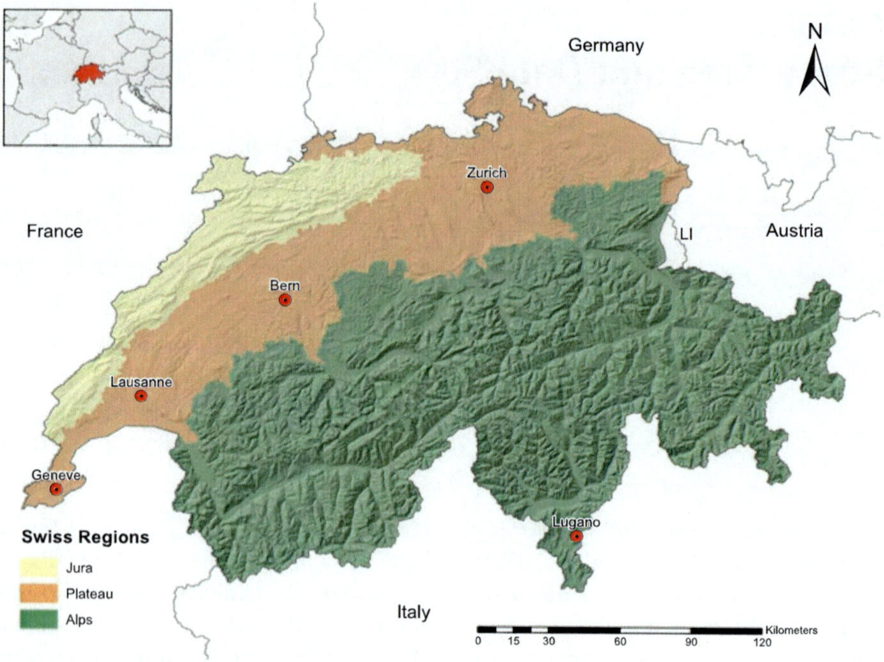

Fig. 2.1 **Study area.** Partition of Switzerland in three geographical regions with topographical relief, adapted from [2]

activities and the majority of the population [2]. As mentioned previously, the wind is highly influenced by terrain features [3]. An example of terrain-forced flows due to topographic characteristics is the channelled wind flow between the Alps and the Jura mountains, which is constrained to circulate on the Plateau [1]. This phenomenon will be referred to hereafter as the *channelling effect*.

In order to describe local topographic shapes, a feature space was constructed and investigated in [3–6] by applying basic image processing operations to a Digital Elevation Model (DEM) of Switzerland of a resolution of $250 \times 250\,[\mathrm{m}^2]$. Table 2.1 presents feature characteristics. In addition to the three features X, Y, Z of the DEM, ten extra features were engineered by considering three quantities potentially relevant for wind speed, namely *Difference of Gaussians* (DoG), *slope* and *directional derivatives*.

These three quantities are based on smoothed versions of the DEM. More precisely, a Gaussian filter with smoothing bandwidth h_0 is used. The directional derivatives in south–north and east–west directions are computed on the smoothed DEM. Remark that these directional derivatives are the two components of the gradient evaluated at each location of the smoothed DEM. By computing the norm of the gradient, one obtains a proportional estimate of the slope at each location of the smoothed DEM. Finally, the DoG is obtained by subtracting two smoothed DEM

Table 2.1 Feature characteristics of the 13-dimensional input space. The ten additional features were engineered using image processing tools to describe local topography [4]

Name	Variable description	Bandwidth parameters
X	X coordinate	
Y	Y coordinate	
Z	Z coordinate (Altitude)	
DoG_s	Difference of Gaussian at small scale	$h_1 = 0.25, h_2 = 0.5$
DoG_m	Difference of Gaussian at medium scale	$h_1 = 1.75, h_2 = 2.25$
DoG_l	Difference of Gaussian at large scale	$h_1 = 3.75, h_2 = 5$
$Slope_s$	Slopes at small scale	$h_0 = 0.2$
$Slope_m$	Slopes at medium scale	$h_0 = 1.75$
$Slope_l$	Slopes at large scale	$h_0 = 3.75$
$DDSN_s$	Dir. deriv. at small scale (south-north)	$h_0 = 0.25$
$DDEW_s$	Dir. deriv. at small scale (east-west)	$h_0 = 0.25$
$DDSN_m$	Dir. deriv. at medium scale (south-north)	$h_0 = 1.75$
$DDEW_m$	Dir. deriv. at medium scale (east-west)	$h_0 = 1.75$

computed with two different bandwidths h_1 and h_2. Note that DoG approximates the Laplacian numerically and is used in edge detection [7]. More details on this feature engineering process can be found in [4].

In terms of topographical and meteorological patterns, DoG detects the ridges, canyons, and depressions, which influences wind speed strength. Slopes could reproduce low wind speeds on the main flanks of big valleys, while the wind speed is higher where the slope vanishes (thalwegs) or even accelerates (e.g. lakes). Directional derivatives are expected to describe wind produced by thermal effects due to sun exposure and other patterns [3].

Ten features are generated by computing the three quantities with different bandwidths, representing terrain characteristics at multiple scales. This data set will be referred to as the *13-dimensional input space*. More details on the building process and meteorological motivations can be found in [3–5].

2.2 MeteoSwiss Wind Speed Data

The data have been provided by and downloaded from the IDAWEB web portal of the Federal Office of Meteorology and Climatology of Switzerland (MeteoSwiss) [8]. These data are gathered by several monitoring networks of ground-level stations. Each monitoring network is measured by different devices and managed by different entities, implying potential heterogeneity in the data. The main monitoring network available in Switzerland—the so-called SwissMetNet—was built between 2003 and 2015 and is handled by MeteoSwiss, which operates several quality checks before making the data available to the users. Its weather stations are geographically well distributed to cover Switzerland and in a manner to represent its different local climates [9]. However, stations are not necessarily representative of the topographic-related features of the 13-dimensional input space.

2.2.1 Data Wrangling, Cleaning and Missing Values

The (scalar) wind speed data consist of 450 stations sampled at 10 [m] above the ground level, every 10 min from 00:00 am the 1st January 2008 to 11:50 pm the 31st December 2017. About 6000 text files were aggregated to form a unique data set. A first glance at the data shows negative values with no physical meaning and are then set to missing values.

Figure 2.2 shows the data availability of wind speed raw data. The top part of the Figure indicates the absence/presence of data for each station along time. As a matter of fact, 48.2% of the data are missing. Some time series have more missing values than available values. For some series, the consecutive lack of data covers several years, which could be partly explained by the progressive introduction of new stations in the monitoring networks. It also seems that some stations have a lower sampling frequency. The number of available stations along time is displayed in the bottom part of Fig. 2.2. Its behaviour—reminding a step function—is possibly due to the commissioning of groups of stations or more likely to the adding of a network in the database. In order to avoid the removal of stations introduced after 2008, the data set is split into three parts:

- from 00:00 am the 1st January 2008 to 11:50 pm the 31st December 2012, which will be referred to as *MSWind 08–12*,
- from 00:00 am the 1st January 2013 to 11:50 pm the 31st December 2016, which will be referred to as *MSWind 13–16*,
- from 00:00 am the 1st January 2017 to 11:50 pm the 31st December 2017, which will be referred to as *MSWind 17*.

For each data set, the stations with more than 10% of missing values are removed. Then, stations are carefully visualised with time series plots, searching for potential outliers that could produce undesirable effects on analysis; see Fig. 2.3 for some examples. All local suspicious behaviours detected by visual inspection have been removed and replaced by missing values. Stations exhibiting global strange behaviour are entirely removed. Moreover, stations with an abnormal amount of zero values were spotted; see Fig. 2.4. Stations with more than 10% of zero values are removed, while the remaining zero values are set to missing values. The frequency of the data is then reduced to 1/hour by averaging. For each data set, the spatial distribution of the remaining stations for each network is visualised in Fig. 2.5. A vast majority of the stations is managed by MeteoSwiss. Stations outside Switzerland are removed (Deutscher Wetterdienst network) because of the non-availability of the 13-dimensional input space outside Switzerland.

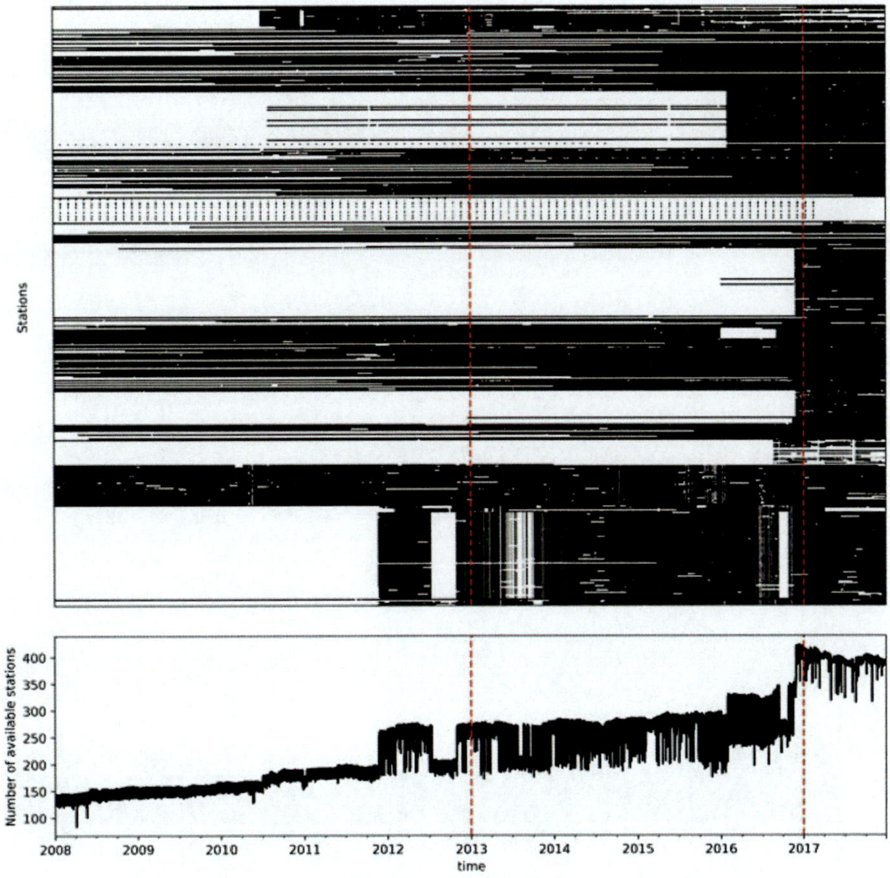

Fig. 2.2 Data availability of the wind speed raw data, from 2008 to 2017: (top) The presence of measurement values is in black, for each of the 450 stations; (bottom) The total number of measurements available. The vertical red dashed lines refer to the splitting of the raw data into three parts

The cleaned data sets are in agreement with the Swiss wind speed records. Switzerland's highest wind speed recorded is 74.4 [m/s] (or 268 [km/h]) in the mountains (Grand St Bernard, the 27th February 1990, due to Hurricane Vivian) and 52.7 [m/s] (190 [km/h]) in the lowlands (Glarus, the 15th July 1985, due to a thunderstorm) [10].

Each of the three data sets is split into a training set and a testing set. More precisely, 80% of the monitoring stations are randomly selected to constitute the training set, while 20% of the remaining stations belong to the testing set. Figure 2.5a and b show that the observation stations are relatively uniformly located in the geographical space and a random splitting process seems appropriate for MSWind 08–12 and MSWind 13–16. However, for MSWind 17, central Switzerland seems slightly

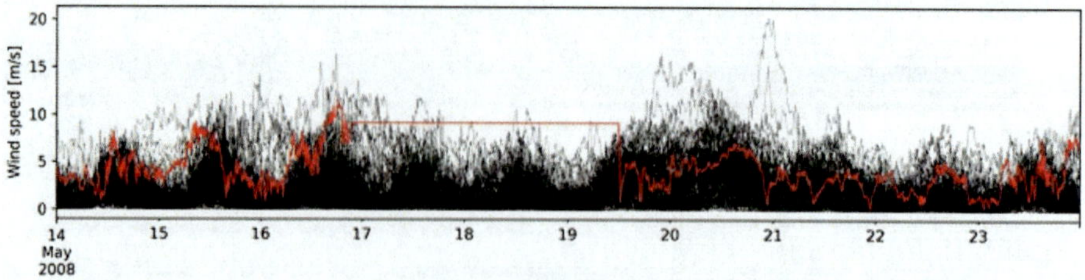

(A) Col du Grand St-Bernard (MeteoSwiss)

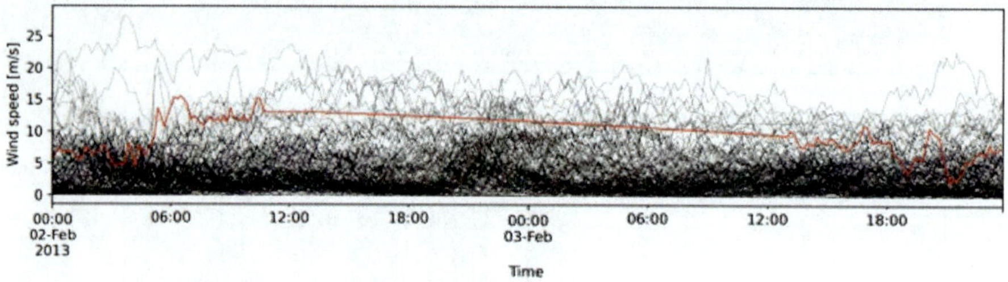

(B) Weissfluhjoch (MeteoSwiss)

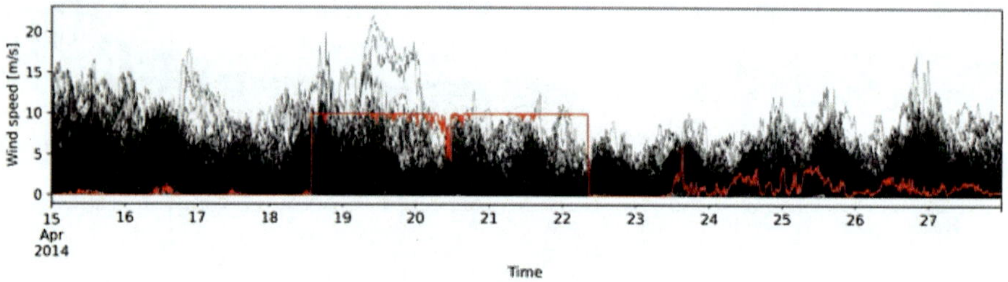

(C) Arosa (MeteoSwiss)

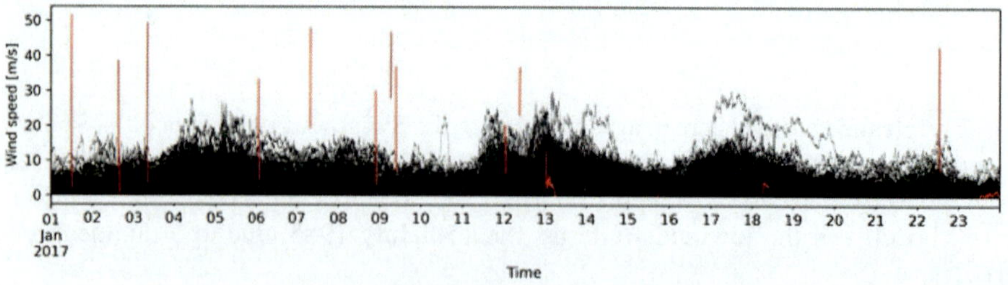

(D) Ufhusen (Kanton Luzern)

Fig. 2.3 Outliers. Some examples of stations with different suspicious behaviours (red) in the wind speed data (black). This kind of behaviour is obviously not characteristic of the wind phenomenon

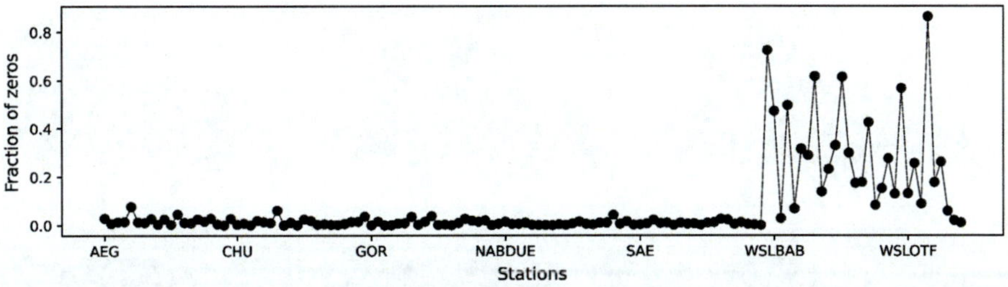

Fig. 2.4 Presence of zero. Fractions of zero values in the wind speed stations for MSWind 08–12

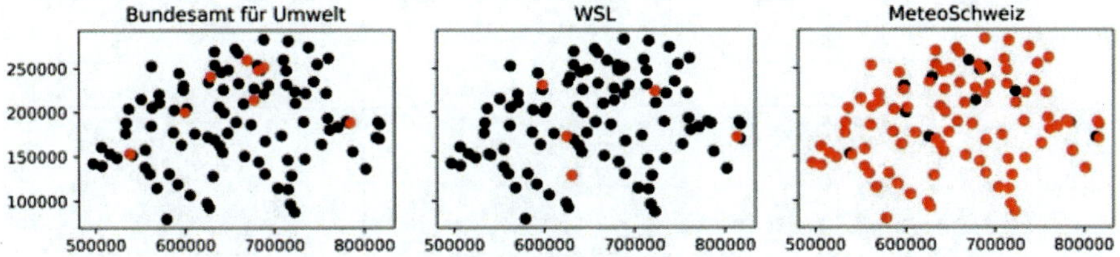

(A) MSWind 08-12

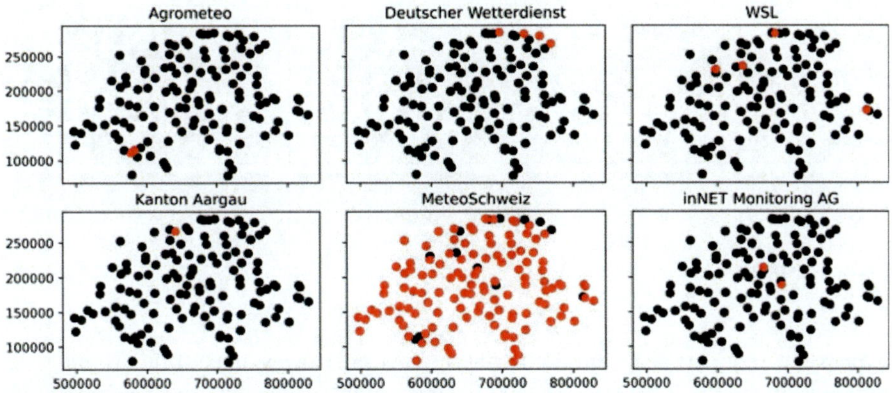

(B) MSWind 13-16

Fig. 2.5 Spatial distribution of each monitoring network. Each network is represented (red) within the wind speed data (black). The Deutscher Wetterdienst network is removed because it does not belong the Swiss territory

over-represented due to the consideration of the Kanton Luzern and Bundesamt für Strassen networks; see Fig. 2.5c. For such a situation, declustering of the monitoring network could be suitable to select a representative testing set of the geographical space [11]. Nevertheless, a random selection is also performed for MSWind 17 to homogenise the methodology between the data sets.

Table 2.2 presents some characteristics of the three cleaned datasets. In particular, for each training set, the percentage of missing values seems now reasonable at the

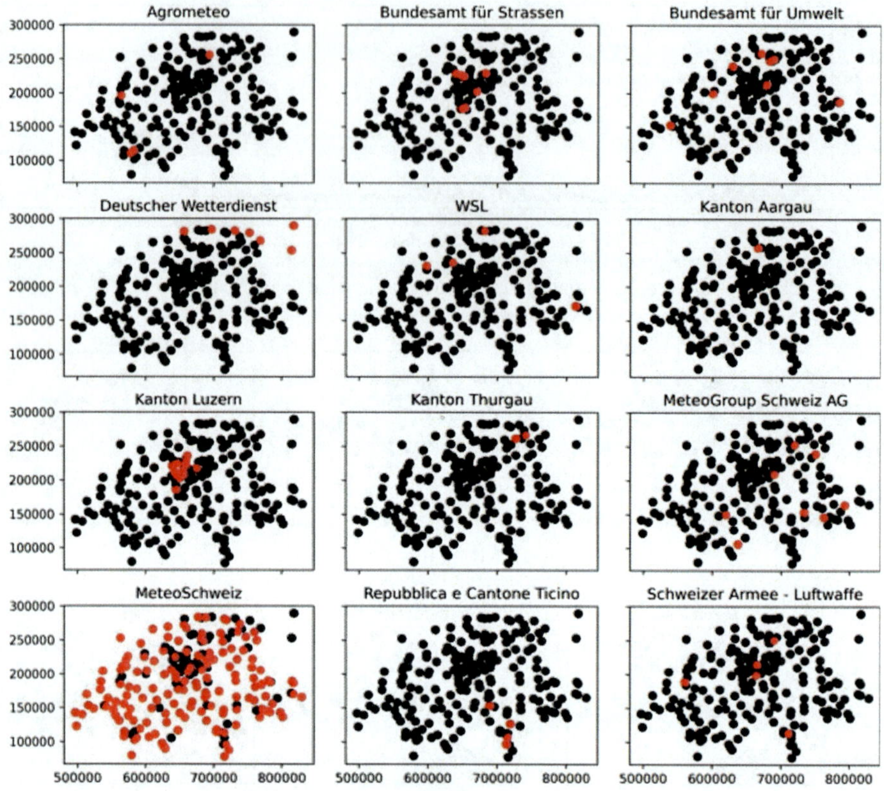

(c) MSWind 17

Fig. 2.5 (continued)

expense of reducing the number of stations. All missing values of the training sets are replaced by the local average data from the eight closer stations in space and the two contiguous time frames, yielding a mean over 24 spatio-temporal neighbours [12, 13].

A Remark on the Splitting

The raw data set was split into three parts to keep a high number of stations during the treatment of the missing values of each part. The choices of doing so, of the number of parts and of the selected points in time used for the splitting are partly motivated by the scientific purpose of investigating the impact of the number of stations on the methodology proposed in Chaps. 5, 6 and 7.

An alternative choice—more oriented to the results of the methodology application—could be to treat missing values on the whole temporal period, keep-

Table 2.2 Cleaned data. Some characteristics of the hourly wind speed data sets after cleaning. When the monitoring network is not clustered, the characteristic network scale can be interpreted as an average distance between the monitoring stations and is computed as $\sqrt{A/n}$ where A is the Switzerland area and n is the number of stations in the network [11]

	MSWind 08–12	MSWind 13–16	MSWind 17
General characteristics			
Total number of stations	106	127	208
Number of training stations	84	101	166
Number of testing stations	22	26	42
Time series length	43'848	35'064	8'760
MeteoSwiss representation	87.7%	92.9%	71.6%
Number of networks	3	5	11
On the training data sets			
Missing values	2.1%	1.1%	1.2%
Minimal distance [km]	≤ 0.1	3.3	0.4
Maximal distance [km]	323.1	332.7	332.7
Characteristic network scale [km]	22.2	20.2	15.8

ing the number of stations probably close to the total number of stations available for the MSWind 08–12 part, and consider them all as training stations. The model could then be tested on the stations removed during the missing value treatment. This would increase the number of training stations compared to MSWind 08–12 and avoid continuity problems at the splitting time points, at the expense of decreasing the number of training stations compared to MSWind 17 and heterogeneous testing along time. However, this alternative choice would give less insight into the methodology assessment.

2.2.2 *Exploratory Data Analysis*

Basic exploratory analysis is more than a mandatory step in any statistical task. Simple tools such as plots and summary statistics provide essential insights into the data. In particular, visualisation must accompany each step of any statistical journey, from raw data to results, including modelling and residuals.

The wind speed spatio-temporal data set—which is hard to visualise as a whole—is explored with time series and spatial plots. For instance, Fig. 2.6 displays several spatial snapshots at different random fixed times for MSWind 13–16. The irregular spatial nature of the data allows us to play with symbol size and colour to describe its values. In some snapshots, structures related to the channelling effect and/or the climatic barrier formed by the Alps chain may somewhat appear.

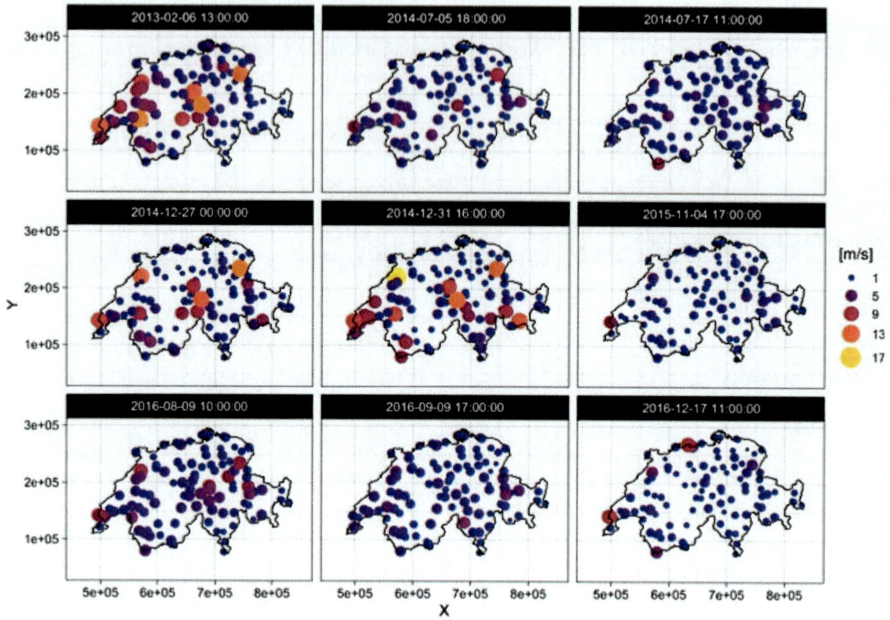

Fig. 2.6 **Spatial snapshots.** Spatial plots at different times for MSWind 13–16. Both colour and dot size describe the wind speed values

For time series plots, fifteen of the MSWind 08–12 monitoring stations are randomly chosen. Their locations and altitudes are reported in Table 2.3. Figure 2.7 shows the time series of hourly wind speed for those stations. Very different behaviours can be observed. Amplitude varies depending on the station and seems not directly related to their altitude. The yearly cycle sometimes seems identifiable, e.g. for the ZER station, while temporal structures are unclear for other stations.

For the same monitoring stations, Autocorrelation Functions (ACFs) against hourly lag are shown in Fig. 2.8. Various structures appear, such as daily cycles with different intensities depending on the station. Sometimes, negative correlations appear at a half-day period, sometimes not at all. Wind speed distributions are approximated with Kernel Density Estimates (KDEs); see Fig. 2.9. KDE will be discussed in more detail in Sect. 4.4. Here, the bandwidth is selected with Scott's rule of thumb [14]. Again, different behaviours can be observed. Some stations seem to have a "classical" behaviour for wind speed distribution, often fitted with a Weibull distribution [15, 16], for instance, NABBER, SCM or TSG. Other stations are more atypical such as AND, EVI, or ENG. The DAV station even seems to exhibit bimodality.

Interesting patterns also appear by plotting the *empirical temporal mean*, i.e. the time series of all the stations averaged across space. An example is given in Fig. 2.10 for MSWind 08–12, with all stations overlayed. Although it suggests a yearly cycle in

Table 2.3 Some monitoring stations of MSWind 08–12. Locations and altitudes for randomly chosen stations used in Figs. 2.7, 2.8 and 2.9

Name	Location	Altitude [m]
AND	Andeer	987
COM	Acquarossa Comprovasco	575
DAV	Davos	1594
DOL	La Dôle	1669
ENG	Engelberg	1035
EVI	Evionnaz	482
FAH	Fahy	596
INT	Interlaken	577
LUZ	Luzern	454
NABBER	Bern	535
NABDUE	Dübendorf	432
SCM	Schmerikon	408
STG	St. Gallen	775
TSG	Arosa (Tschuggen)	2040
ZER	Zermatt	1638

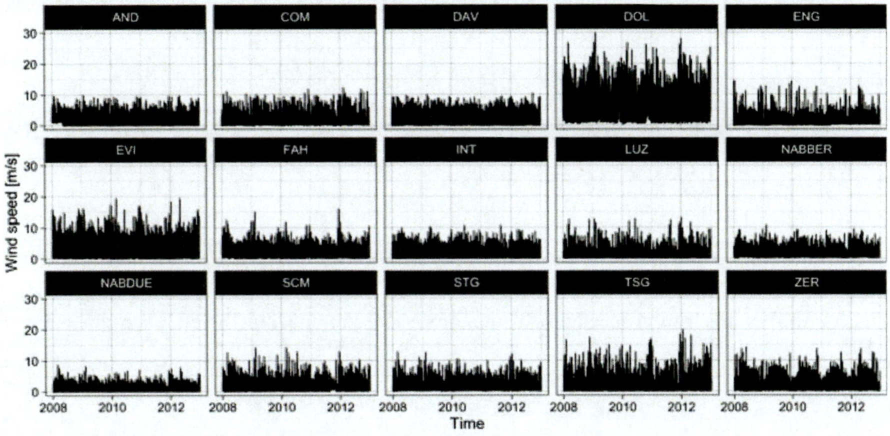

Fig. 2.7 Time series plots. Wind speed measurements of stations presented in Table 2.3

the empirical temporal mean variability—which seems slightly higher in the winter—it is hard to see local patterns at such temporal scale, and some closer plots are provided for different months in Fig. 2.11. For June 2008 (2.11a) and slightly less for April 2015 (2.11c), a clear daily cycle is present. The daily cycle is alleviated for October 2013 (2.11b) and completely disappears for January 2017 (2.11d).

The monitoring station representativity was also investigated. Although the stations are spatially well represented, this is not necessarily the case in the 13-

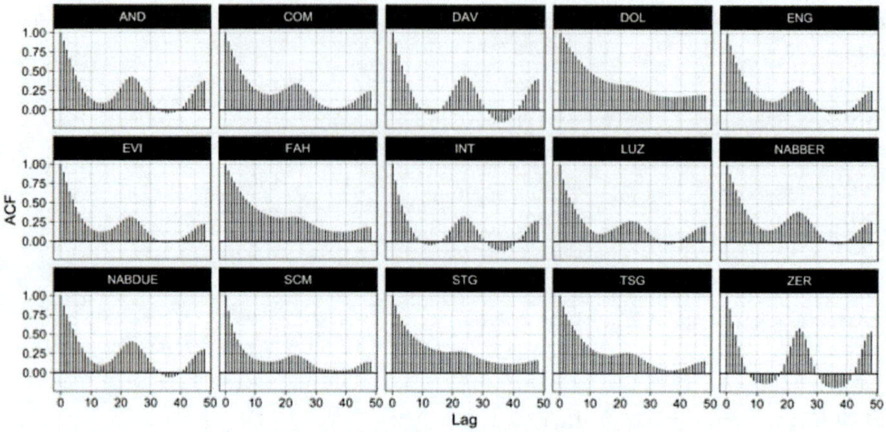

Fig. 2.8 ACFs. Wind speed ACF of each station presented in Table 2.3

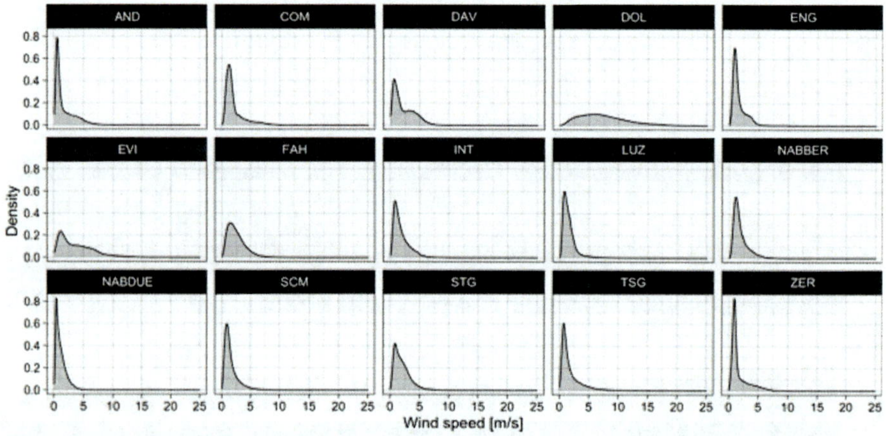

Fig. 2.9 KDEs. Each wind speed distribution of stations presented in Table 2.3 is estimated with KDE

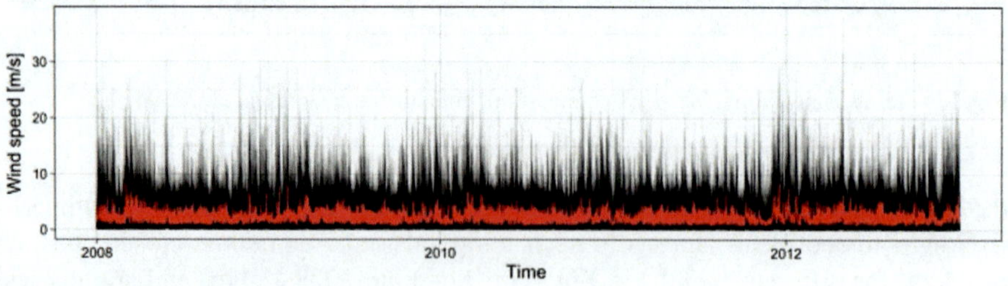

Fig. 2.10 Average of all stations. In red, the empirical temporal mean times series for MSWind 08–12. Single stations are also displayed in black

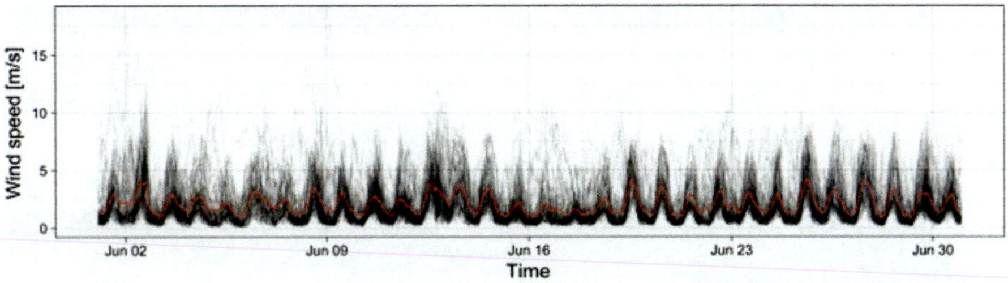

(A) June 2008

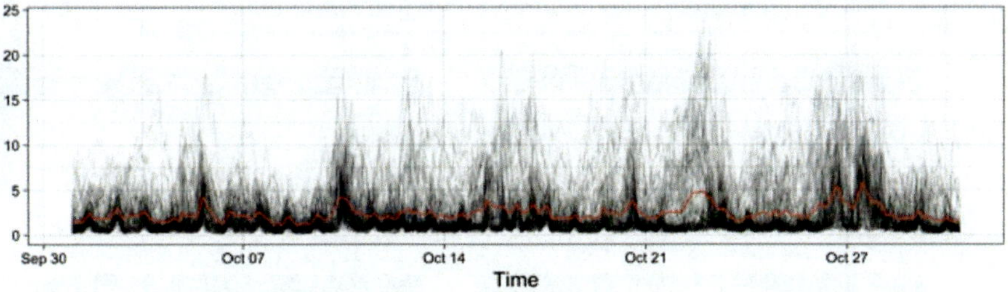

(B) October 2013

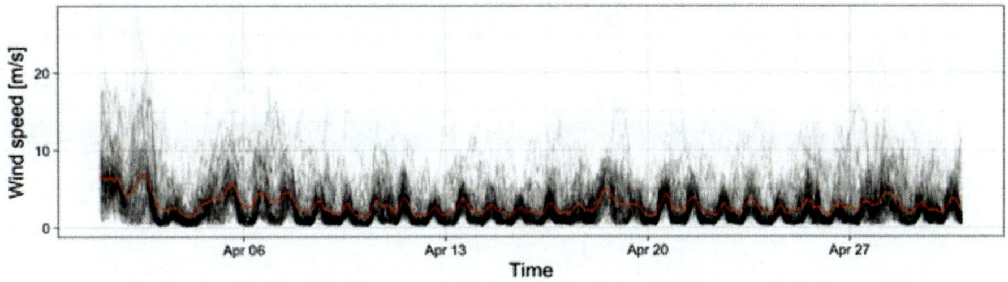

(C) April 2015

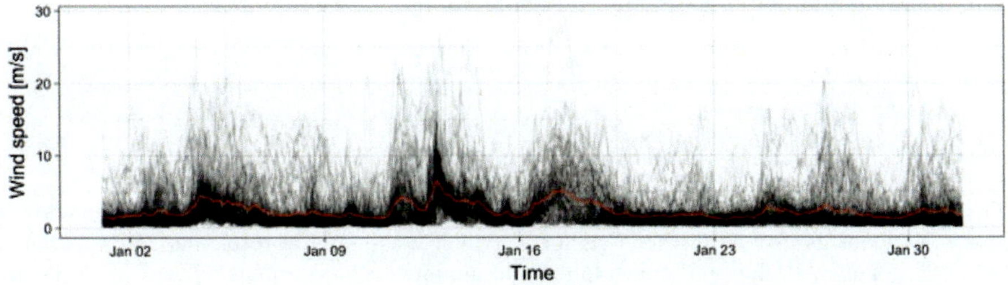

(D) January 2017

Fig. 2.11 Average of all stations—zoom in different months. In red, the empirical temporal mean times series for different months. Single stations are also displayed in black

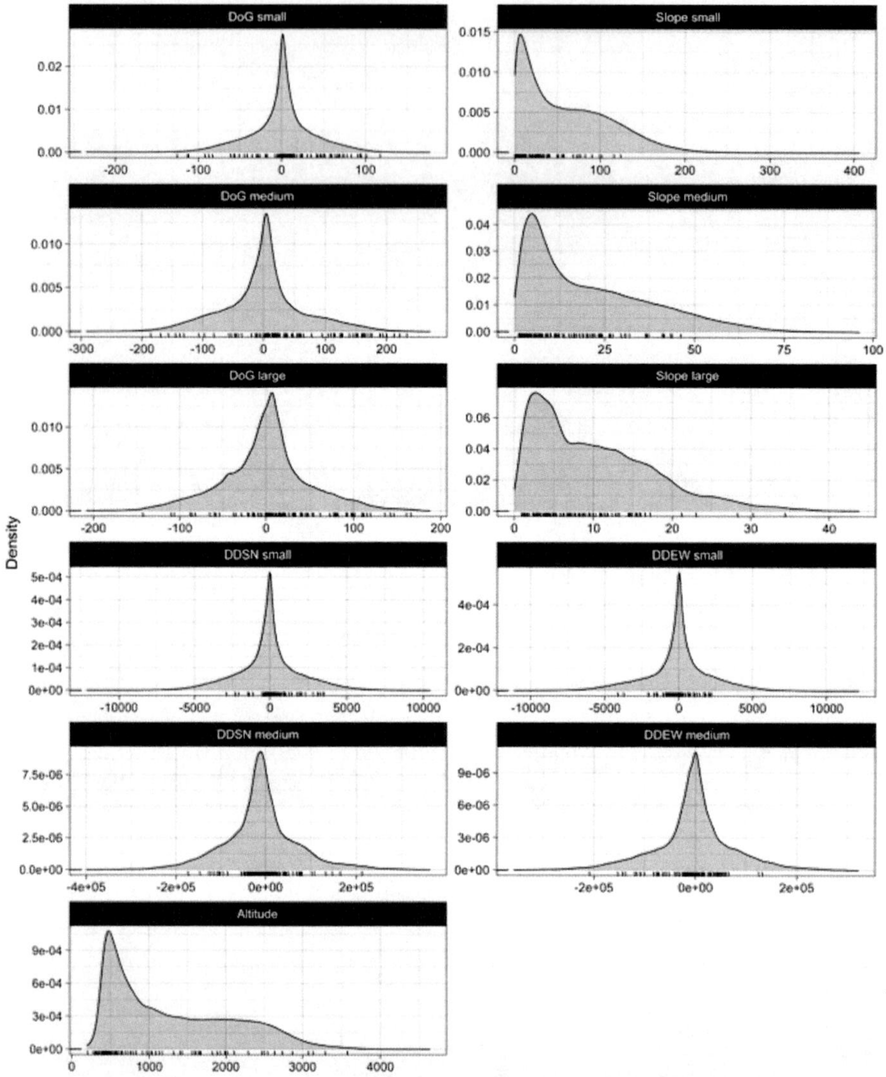

Fig. 2.12 Representativity of the MSWind 13–16 training data set in the 13-dimensional input space. KDEs of the feature distributions for the whole Swiss grid of resolution 250 × 250 [m^2] (660’697 points). All plots are limited to the grid minimum and maximum values. The MSWind 13–16 training stations are reported as marks under the densities (101 points). Coordinates X and Y are not shown

dimensional input space. Figure 2.12 shows KDEs of the topographic-related features computed from the whole grid presented in Sect. 2.1. The 101 training stations of MSWind 13–16 are marked below the densities as ticks. The altitude density plot shows that most monitoring stations are below 800 [m], and far less information on the phenomena is available for the higher elevations. Features such as small and medium slopes have even no sampling points for their right tails. Directional derivatives are also mainly sampled around the mean. In a modelling context, it is hard to say what happened in the high dimensional input space. For instance, interpolation in the geographical space may lead to extrapolation in the high-dimensional space. This partly motivates the interest in UQ.

2.3 MoTUS High-Frequency Wind Speed Data

The MoTUS (Measurement of Turbulence in an Urban Setup) is an experiment that aims to increase the understanding of vertical wind profile, temperature and turbulence in an urban setup. It consists of a vertical mast of 27 [m] in height installed on the campus of Ecole Polytechnique Fédérale de Lausanne (EPFL) [17]; see Figs. 2.1 and 2.13. The average building height is around 10 [m] [18, 19]. Seven 3D sonic anemometers (Gill Wind Master) are located every 4 m on the mast. The anemometers were set at 1.5, 5.5, 9.5, 13.5., 17.5, 21.5 and 25.5 m above the ground. The highest anemometer was placed far enough above the building layout height to be in an undisturbed flow [20].

One goal of this experiment is to study the impact of the surrounding building on the wind profile. Indeed, the building layout could be considered as the dominant source of local turbulence phenomena in the wind flow below the average height of the buildings.

An extremely interesting feature of this monitoring station is the very high resolution of its measurements. The three velocity components of the three-dimensional wind vector, the sonic speed and sonic temperature are recorded at a frequency of 20 [Hz] for each level on the tower. As a matter of fact, a slight variation in the sampling frequency appears in the raw data. In order to obtain a regular sampling frequency, the data have been averaged at 1 [Hz]. The atmospheric pressure is measured on the site at a 1 [Hz] frequency, with a Gill meteorological station (GMX300). These data have been provided by Dr. Dasaraden Mauree from the Solar Energy and Building Physics Laboratory (LESO-PB) of the EPFL, who conducted the experiment.

2.3.1 *Data Wrangling, Cleaning and Missing Values*

The raw data cover about five months of measurements which corresponds to hundreds of millions of entries. After formatting correction and files aggregation, wind speed—computed as the euclidean norm of the three velocity components—sonic

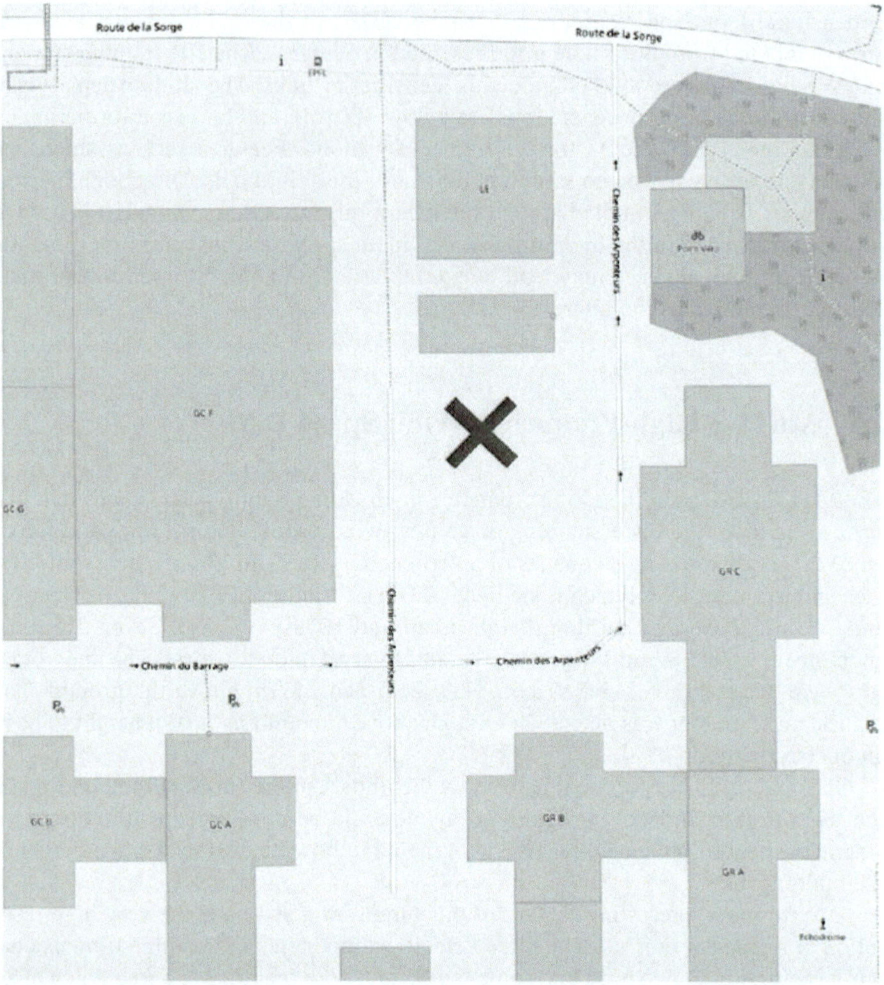

Fig. 2.13 Emplacement of the mast. This image is taken and modified from Open Street Map, whose copyright notices can be found at https://www.openstreetmap.org/copyright

temperature and pressure were extracted. About two months of data are selected, in a manner to minimise the number of consecutive missing values. Indeed, failures in one of the sensors—even a short duration—generate many missing values at such frequency. The remaining sparse missing values on this period represent about 0.02% and are replaced by local linear interpolation in time. The selected period, from 28 November 2016 to 29 January 2017, will be referred to as the *MoTUS* dataset in the remainder of the book.

Table 2.4 Summary statistics for the MoTUS wind speed data. Statistics are computed for each anemometer. Results are in [m/s]

	An 1	An 2	An 3	An 4	An 5	An 6	An 7
Height [m]	1.5	5.5	9.5	13.5	17.5	21.5	25.5
Min.	0.000	0.000	0.000	0.000	0.000	0.000	0.010
1st Qu.	0.278	0.351	0.481	0.920	1.173	1.280	1.395
Median	0.493	0.602	0.824	1.575	1.965	2.124	2.295
Mean	0.606	0.721	1.009	1.932	2.388	2.574	2.756
3rd Qu.	0.812	0.955	1.324	2.603	3.200	3.435	3.661
Max.	7.774	12.254	14.659	18.583	20.397	21.611	23.010

2.3.2 Exploratory Data Analysis

The wind speed observed by the higher anemometers would be in conditions of undisturbed flow.[1] By contrast, the three lowest sensors would be more impacted by the canopying and tunnelling effect produced by the layout of the buildings around the mast. Indeed, the surface heterogeneity in the urban canyon—grass, ground, trees, building surfaces—generates additional sources of turbulence leading to a more chaotic flow [21] and breaking down the larger turbulent eddies into smaller ones [22–24], which is also in agreement with previous studies such as [23, 25, 26].

Summary statistics for the seven anemometers are given in Table 2.4. Each anemometer is labelled from 1 (1.5 [m]) to 7 (25.5 [m]). The high-frequency wind speed time series for the two months are shown in Fig. 2.14. It can be seen that all the time series seem to exhibit a similar variability structure, although the amplitude of the variation increases with the height from the ground. This is confirmed by computing the Pearson correlation, see Table 2.5, which shows correlation coefficients ranging from 0.55 to 0.96.

Although such similarity is observed among the series through correlation coefficients, the wind speed series are governed by different local forcings characterising the variability. Actually, the wind speed variability up to 9.5 [m]—slightly below the average building height around the mast—should be more informative on the small turbulence dynamics happening around buildings, which is characteristic of the roughness sublayer [27, 28]. The wind speed variability from 17.5 [m] up to 25.5 [m]—so, sufficiently above the average building layout elevation—would be in unconditioned flow status, characteristic of the inertial layer, where no significant differences in the turbulent fluxes along the vertical direction take place [27, 28]. The wind speed observed at 13.5 [m] is likely in an intermediate status and thus may be affected by both phenomena. Particularly at this height, the rooftop may influence the turbulent structure creating more disturbed flow above the building height [29]. Since the 13.5 [m] anemometer is positioned at the interface between the inertial sub-layer

[1] In this subsection, physical interpretations were provided by Dr. Dasaraden Mauree and Dr. Luciano Telesca.

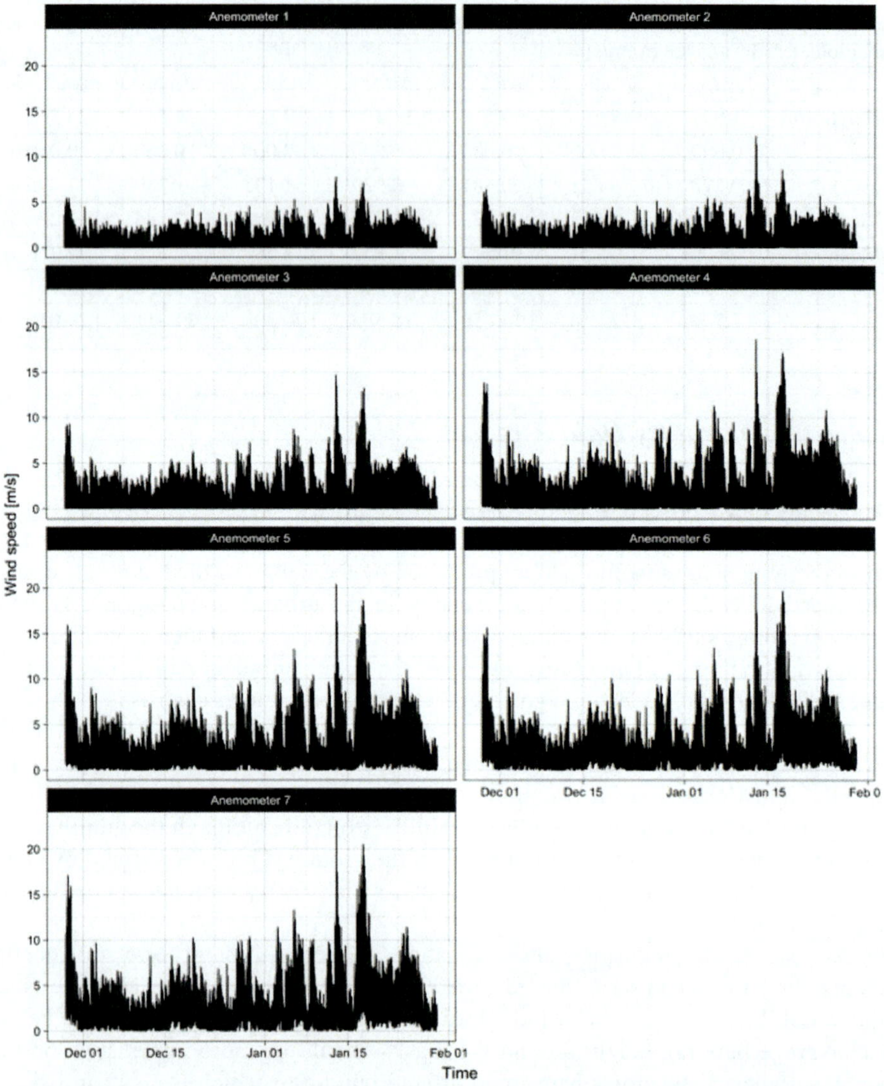

Fig. 2.14 1 Hz wind speed time series for the seven MoTUS anemometers. All the time series seem to show a similar variability structure, with an amplitude of the variation increasing with altitude

and the urban canopy layer, it will probably show the characteristics of both layers. Over and above this, wake turbulence can be described as supplementary turbulent flows arising from the shape of the obstacles in the wake of the obstacles [30]. Therefore, this could lead to higher turbulent activities in this particular region. Moreover, wake turbulence may also be responsible for a higher variance at this height [31].

Table 2.5 Pearson correlations between anemometers. All pairs of time series present important correlations

	An 1	An 2	An 3	An 4	An 5	An 6	An 7
An 1	1.000	0.641	0.550	0.600	0.628	0.634	0.637
An 2	0.641	1.000	0.620	0.574	0.590	0.597	0.601
An 3	0.550	0.620	1.000	0.677	0.625	0.616	0.612
An 4	0.600	0.574	0.677	1.000	0.895	0.854	0.830
An 5	0.628	0.590	0.625	0.895	1.000	0.947	0.915
An 6	0.634	0.597	0.616	0.854	0.947	1.000	0.959
An 7	0.637	0.601	0.612	0.830	0.915	0.959	1.000

Variability and turbulence of these data will be explored in Sect. 3.3 with the wavelet variance, a tool designed to describe the variance depending on the scale.

KDEs of the high-frequency wind speed time series are shown in Fig. 2.15. Independently of the increasing variance with altitude, the three highest anemometers' density shape seems different from the three lowest density. This aspect will be further investigated in Chap. 4 with the Fisher-Shannon analysis, a multi-use tool including the possibility of discriminating distributions of different shapes.

For visualisation purposes, all density plots are limited to a wind speed of 10 [m/s]. However, a certain amount of extreme values—typical of wind data—are present, as shown by boxplots in Fig. 2.16.

2.4 Summary

A 13-dimensional input space was presented, which enriched the geographical space with engineered features related to the local geometric properties of the terrain. Those features are expected to help the learning of the spatio-temporal wind speed field in Chap. 7 of this book. However, training data do not necessarily represent the prediction range in this 13-dimensional input space and could lead to extrapolation, even in interpolation situations in the geographical space. This constitutes an additional motivation for considering reliable uncertainty assessments of the prediction, which will be developed for ELM in Chap. 6 and should detect and quantify such representativity issues.

The basic EDA already demonstrate a rich and complex picture of the MeteoSwiss and MoTUS data. It also suggests the difficulty of extracting relevant information and effective summaries from such spatio-temporal wind speed data sets. Advanced exploratory tools will be presented and used in Chaps. 3 and 4 to gain more insights.

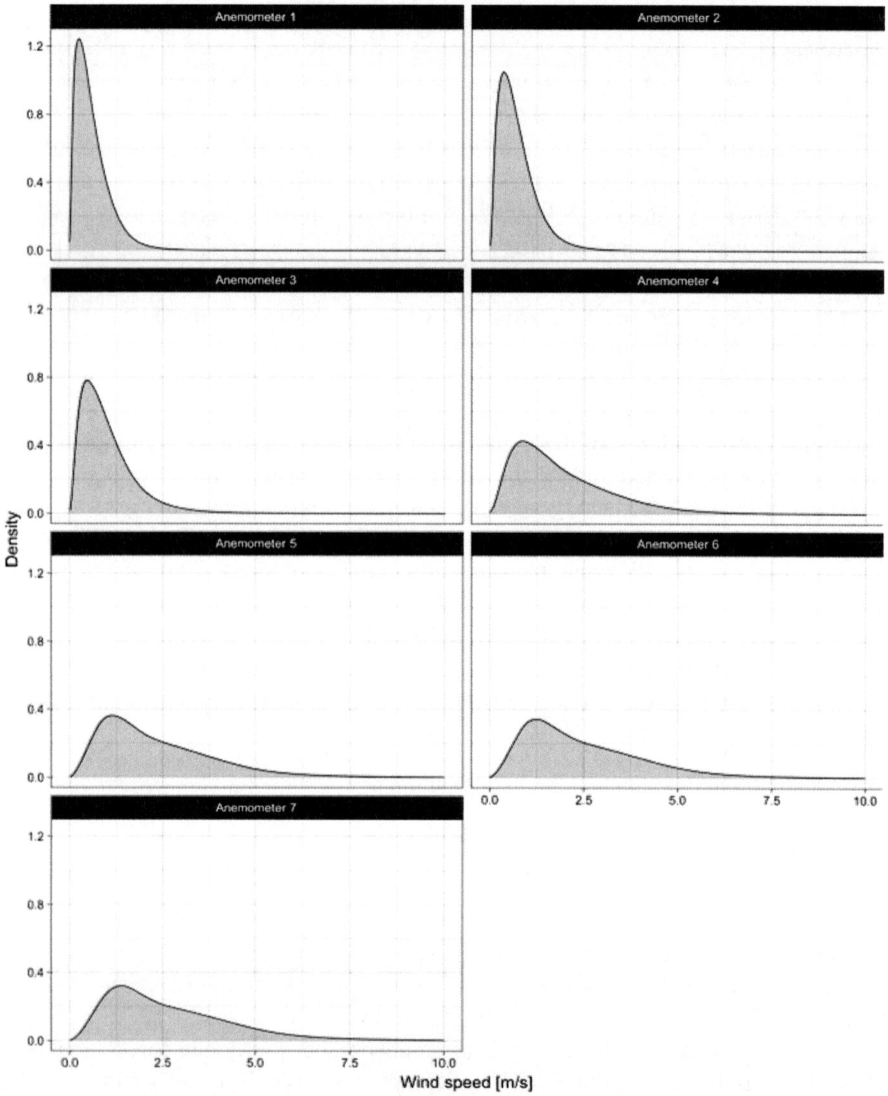

Fig. 2.15 KDEs of the wind speed distribution for the seven MoTUS anemometers. All plots are limited to a wind speed of 10 [m/s] for visualisation purposes

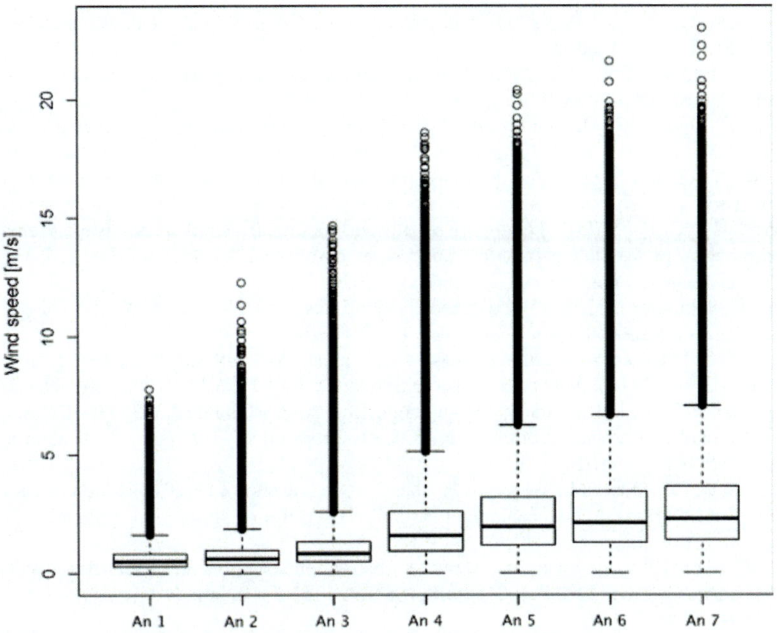

Fig. 2.16 Boxplots for the seven MoTUS anemometers. An important amount of extreme values are present, typical of wind speed data

References

1. Whiteman C (2000) Mountain Meteorology: fundamentals and applications. Oxford University Press, Oxford
2. Vega Orozco CD, Golay J, Kanevski M (2015) Multifractal portrayal of the swiss population. Cybergeo: Eur J Geogr 714(1):1–19. https://doi.org/10.4000/cybergeo.26829
3. Robert S, Foresti L, Kanevski M (2013) Spatial prediction of monthly wind speeds in complex terrain with adaptive general regression neural networks. Int J Climatol 33(7):1793–1804
4. Foresti L (2011) Kernel-based mapping of meteorological fields in complex orography. Ph.D. Dissertation, UNIL Lausanne
5. Foresti L, Tuia D, Kanevski M, Pozdnoukhov A (2011) Learning wind fields with multiple kernels. Stoch Envir Res Risk Assess 25(1):51–66
6. Laib M, Kanevski M (2019) A new algorithm for redundancy minimisation in geo-environmental data. Comput & Geosci 133(104):328
7. Marr D, Hildreth E (1980) Theory of edge detection. Proc R Soc Lond Ser B Biolog Sci 207(1167):187–217
8. Meteoswiss (Federal Office of Meteorology and Climatology of Switzerland), Idaweb 1.3.5.0 (2016). https://gate.meteoswiss.ch/idaweb
9. Meteoswiss (Federal Office of Meteorology and Climatology of Switzerland), Swiss-metnet: the meteoswiss reference monitoring network (2018). https://www.meteoswiss.admin.ch/content/dam/meteoswiss/en/Mess-Prognosesysteme/Bodenstationen/doc/SwissMetNetTheMeteoSwissReferenceMonitoringNetwork.pdf. Accessed 23 Oct 2020
10. Meteoswiss (Federal Office of Meteorology and Climatology of Switzerland), Swiss records (2020). https://www.meteoswiss.admin.ch/home/climate/the-climate-of-switzerland/rekorde-schweiz.html. Accessed 23 Oct 2020

11. Kanevski M, Maignan M (2004) Analysis and modelling of spatial environmental data, vol 6501. EPFL Press, Lausanne
12. Jun M, Stein ML (2007) An approach to producing space-time covariance functions on spheres. Technometrics 49(4):468–479
13. Porcu E, Bevilacqua M, Genton MG (2016) Spatio-temporal covariance and cross-covariance functions of the great circle distance on a sphere. J Amer Stat Assoc 111(514):888–898
14. Scott DW (2015) Multivariate density estimation: theory, practice, and visualization. Wiley, New York
15. Veronesi F, Grassi S (2015) Comparison of hourly and daily wind speed observations for the computation of Weibull parameters and power output. In: 3rd International renewable and sustainable energy conference (IR- SEC). IEEE 2015, pp 1–6
16. Jung C, Schindler D (2019) Wind speed distribution selection-a review of recent development and progress. Renew Sustain Energy Rev 114(109):290
17. EPFL (2021) Urban microclimate measurement mast - MoTUS. http://motus.epfl.ch
18. Mauree D, Lee DS-H, Naboni E, Coccolo S, Scartezzini J-L (2017) Localized meteorological variables influence at the early design stage. Energy Procedia 122:325–330. CISBAT 2017 international conference, future buildings & districts - energy efficiency from nano to urban scale. ISSN: 1876-6102
19. Mauree D, Deschamps L, Bequelin P, Loesch P, Scartezzini J-L (2017) Measurement of the impact of buildings on meteorological variables. In: Building simulation application proceedings. Bu Press, Bolzano
20. Rotach MW (1999) On the influence of the urban roughness sublayer on turbulence and dispersion. Atmosph Envir 33(24):4001–4008. ISSNN: 1352-2310. https://doi.org/10.1016/S1352-2310(99)00141-7
21. Zaïdi H, Dupont E, Milliez M, Musson-Genon L, Carissimo B (2013) Numerical simulations of the microscale heterogeneities of turbulence observed on a complex site. Bound-Layer Meteorol 147(2):237–259. https://doi.org/10.1007/s10546-012-9783-9
22. Coceal O, Belcher SE (2004) A canopy model of mean winds through urban areas. Q J R Meteorol Soc 130:1349–1372. https://doi.org/10.1256/qj.03.40
23. Mauree D, Blond N, Kohler M, Clappier A (2017) On the coherence in the boundary layer: development of a canopy interface model. Front Earth Sci 4:109. ISSN: 2296-6463. https://doi.org/10.3389/feart.2016.00109
24. Raupach MR, Finnigan JJ, Brunei Y (1996) Coherent eddies and turbulence in vegetation canopies: the mixing-layer analogy. Bound-Layer Meteorol 78(3):351–382. https://doi.org/10.1007/BF00120941
25. Santiago JL, Martilli A, Martín F (2007) Cfd simulation of airflow over a regular array of cubes. part i: three-dimensional simulation of the flow and validation with wind-tunnel measurements. Bound-Layer Meteorol 122(3):609–634. ISSN: 1573-1472. https://doi.org/10.1007/s10546-006-9123-z
26. Christen A, Rotach MW, Vogt R (2009) The budget of turbulent kinetic energy in the urban roughness sublayer. Bound-Layer Meteorol 131:193–222. https://doi.org/10.1007/s10546-009-9359-5
27. Fisher P, Kukkonen J, Piringer M, W RM, Schatzmann M (2005) Meteorology applied to urban air pollution problems: Concepts from cost 715. Atmosph Chem Phys Dis Eur Geosci Union 5:7903–7927
28. Oke T (1976) The distinction between canopy and boundary-layer urban heat islands. Atmosphere 14(4):268–277. https://doi.org/10.1080/00046973.1976.9648422
29. Coceal O, Dobre A, Thomas TG, Belcher SE (2007) Structure of turbulent flow over regular arrays of cubical roughness. J Fluid Mech 589:375–409. https://doi.org/10.1017/S002211200700794X
30. Oke TR, Mills G, Christen A, Voogt J (2017) Urban climates. Cambridge University Press, Cambridge
31. Oke T (1988) Street design and urban canopy layer climate. Energy Build 11(1):103–113. ISSN: 0378-7788. https://doi.org/10.1016/0378-7788(88)90026-6

Chapter 3
Advanced Exploratory Data Analysis

This chapter introduces some advanced EDA tools which can be applied to spatio-temporal data sets. They quantify and confirm some features of the wind speed data suggested by visualisation in the previous chapter and unveil hidden patterns. Section 3.1 discusses Empirical Orthogonal Functions (EOFs), a well-known method to analyse spatio-temporal fields. Section 3.2 presents variography, another popular tool coming from geostatistics used to describe spatio-temporal linear dependencies. Section 3.3 deals with wavelet variance, a signal processing technique that allows exploring signal variance at multiple scales.

3.1 Empirical Orthogonal Functions

EOF analysis is a popular approach in climatology, meteorology, oceanography [1] and studying the variability in a geophysical field of interest [2]. In its most common formulation—the spatial one, see the following subsection—EOF analysis can be seen as an application of PCA in the case where data is a multivariate spatially indexed vector with multiple samples over time [2, 3].

3.1.1 Spatial Formulation

Let us suppose that one has spatio-temporal observations $\{Z(\mathbf{s}_i, t_j)\}$ at S spatial locations $\{\mathbf{s}_i : 1 \leq i \leq S\}$ and T time-indices $\{t_j : 1 \leq j \leq T\}$, with $S \leq T$. Let $\widetilde{Z}_s(\mathbf{s}_i, t_j)$ be the *spatially centred data*,

$$\widetilde{Z}_s(\mathbf{s}_i, t_j) := Z(\mathbf{s}_i, t_j) - \widehat{\mu}_s(\mathbf{s}_i),$$

F. Guignard, *On Spatio-Temporal Data Modelling and Uncertainty Quantification Using Machine Learning and Information Theory* , Springer Theses,
https://doi.org/10.1007/978-3-030-95231-0_3

where

$$\widehat{\mu}_s(\mathbf{s}_i) := \frac{1}{T}\sum_{j=1}^{T} Z(\mathbf{s}_i, t_j),$$

is the *empirical spatial mean* at the location $\mathbf{s}_i$. Then the centred data $\widetilde{Z}_s(\mathbf{s}_i, t_j)$ can be represented with a discrete spatial orthonormal basis $\{\phi_k(\mathbf{s}_i)\}_{k=1}^{S}$, i.e.

$$\widetilde{Z}_s(\mathbf{s}_i, t_j) = \sum_{k=1}^{S} a_k(t_j)\phi_k(\mathbf{s}_i), \tag{3.1}$$

such that

$$\begin{aligned}
&\mathbb{E}[a_k(t_j)] = 0, \text{ for } k = 1, \dots S, \\
&\mathrm{Var}[a_1(t_j)] \geq \mathrm{Var}[a_2(t_j)] \geq \cdots \geq \mathrm{Var}[a_S(t_j)] \geq 0, \\
&\mathrm{Cov}[a_k(t_j), a_l(t_j)] = 0, \text{ for all } k \neq l,
\end{aligned}$$

where $a_k(t_j)$ is the coefficient with respect to the kth basis function ϕ_k at time t_j, and $\mathbb{E}[\,\cdot\,]$, $\mathrm{Var}[\,\cdot\,]$, $\mathrm{Cov}[\,\cdot\,,\,\cdot\,]$ denote respectively the expectation, variance and covariance. Notice that the scalar coefficient $a_k(t_j)$ only depends on the time and not on the location, while the spatial basis function $\phi_k(\mathbf{s}_i)$ is independent of time.

Provided the invariance in time of the spatial covariance function of $\widetilde{Z}_s$ [4], the decomposition (3.1) is theoretically justified by the Karhunen-Loève expansion [5, 6], based on Mercer's theorem [7].

Let us introduce some matrix notations. The space-wide data matrix $\mathbf{Z} \in \mathbb{R}^{T\times S}$ is defined element-wise by $\mathbf{Z}_{ji} = Z(\mathbf{s}_i, t_j)$. The jth row of $\mathbf{Z}$ is the S-dimensional vector denoted by $\mathbf{Z}_{t_j}$ and contains the observations at all locations at time t_j. Then, the empirical spatial mean can be written as a vector $\widehat{\boldsymbol{\mu}}_s \in \mathbb{R}^S$, with

$$\widehat{\boldsymbol{\mu}}_s := \frac{1}{T}\sum_{j=1}^{T}\mathbf{Z}_{t_j}.$$

The space-wide centred data matrix is defined by $\widetilde{\mathbf{Z}} := \mathbf{Z} - \mathbb{1}\,\widehat{\boldsymbol{\mu}}_s^T$, where $\mathbb{1}$ is a vector in which all elements are equal to 1 and $(\,\cdot\,)^T$ denotes the transpose operator. Although $\widetilde{\mathbf{Z}}$ is the matrix of the spatially centred data $\widetilde{Z}_s(\mathbf{s}_i, t_j)$, the "$s$" subscript is dropped to lighten the matrix notation.

The basis computation is related to the spectral decomposition of the *empirical spatial covariance matrix* [8], which can be computed as

$$\widehat{\mathbf{C}}_s := \frac{1}{T-1}\sum_{j=1}^{T}(\mathbf{Z}_{t_j} - \widehat{\boldsymbol{\mu}}_s)(\mathbf{Z}_{t_j} - \widehat{\boldsymbol{\mu}}_s)^T = \frac{1}{T-1}\widetilde{\mathbf{Z}}^T\widetilde{\mathbf{Z}}, \tag{3.2}$$

assuming that the variance of $\mathbf{Z}_{t_j}$ does not depend on time. As this real matrix is symmetric and non-negative definite, the following spectral decomposition holds,

$$\widetilde{\mathbf{Z}}^T\widetilde{\mathbf{Z}} = \mathbf{\Phi}^T \mathbf{\Lambda} \mathbf{\Phi}, \tag{3.3}$$

where $\mathbf{\Lambda} = \text{diag}(\lambda_1, \ldots, \lambda_S)$ is the diagonal matrix of the non-negative eigenvalues decreasing down the diagonal, and $\mathbf{\Phi}^T = [\phi_1|\ldots|\phi_S]$ is the orthogonal matrix of the corresponding spatially indexed eigenvectors $\phi_k = (\phi_k(\mathbf{s}_1), \ldots, \phi_k(\mathbf{s}_S))^T$, for $k = 1, \ldots, S$, also called *Empirical Orthogonal Functions* (EOFs).

The EOFs form a discrete orthonormal basis, as $\mathbf{\Phi}^T\mathbf{\Phi} = \mathbf{\Phi}\mathbf{\Phi}^T = \mathbf{I}$, where $\mathbf{I}$ is the identity matrix [4]. The kth *Principal Component* (PC) *time series* is the time series of coefficients of the corresponding EOF, or equivalently the contribution of the kth spatial basis at time t_j, and is then given by $a_k(t_j) = (\mathbf{Z}_{t_j} - \widehat{\boldsymbol{\mu}}_s)^T \phi_k$, $j = 1, \ldots, T$ [8]. By noting $\mathbf{A} \in \mathbb{R}^{T\times S}$ the matrix defined by $\mathbf{A}_{jk} = a_k(t_j)$, this can be written $\mathbf{A} = \widetilde{\mathbf{Z}}\mathbf{\Phi}^T$, and the PC time series are the columns of this matrix. Hence,

$$\widetilde{\mathbf{Z}} = \widetilde{\mathbf{Z}}\mathbf{\Phi}^T\mathbf{\Phi} = \mathbf{A}\mathbf{\Phi}$$

which is the matrix formulation of Eq. (3.1).

Practically, EOFs computation is equivalent to a PCA [9–11], where space-indices are considered as variables and the realisations of those variables correspond to the temporal realisations of the phenomenon. As in classical PCA, the relative variance of each basis is given by the corresponding eigenvalue. Therefore, in the case of reconstruction with a truncated number $K \leq S$ of components, the decomposition (3.1) compresses the spatio-temporal data and reduces its noise. Note also that, although it was assumed that $S \leq T$, the EOFs can also be calculated for the case $S > T$; see [4] for technical details.

Figure 3.1 displays an application of EOF decomposition on the MSWind 13–16 data set.[1] The first three components explain about 52% of the variation in the data. The spatial basis elements (3.1a) with their corresponding normalised PC time series (3.1b) shows the emergence of interesting patterns. Note that the normalised PC times series are the PC times series divided by their standard deviation, which are the columns of the matrix $\mathbf{A}\mathbf{\Lambda}^{-1/2}$. The normalised PC time series are zoomed-in April 2015 to have a closer look (3.1c). The first spatial EOF shows negative values on the whole of Switzerland with a slightly higher magnitude at some known windy stations. Its corresponding temporal coefficients show the contribution of this spatial pattern in time, which is seen to be more variable during the winter. The second spatial EOF shows a clear spatial pattern that splits the country in two, and the Alpine region seems recognisable. The spatial pattern of the third EOF is harder to interpret, although some windy stations seem to stand out as in the first EOF. However, the corresponding temporal coefficients exhibit a clear yearly cycle, while a closer look in April 2015 shows a distinct daily cycle.

[1] Inspired from the R code of [8].

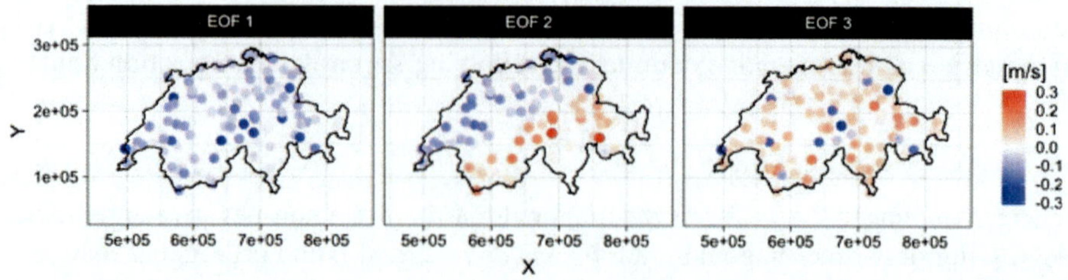

(A) Spatial EOF basis elements.

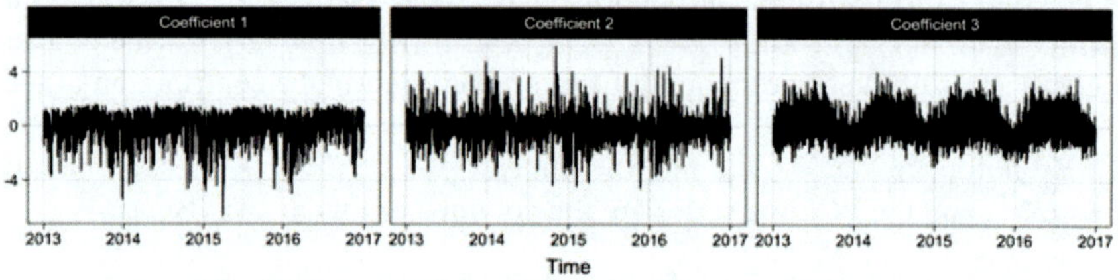

(B) Normalised PC time series acting as temporal coefficients.

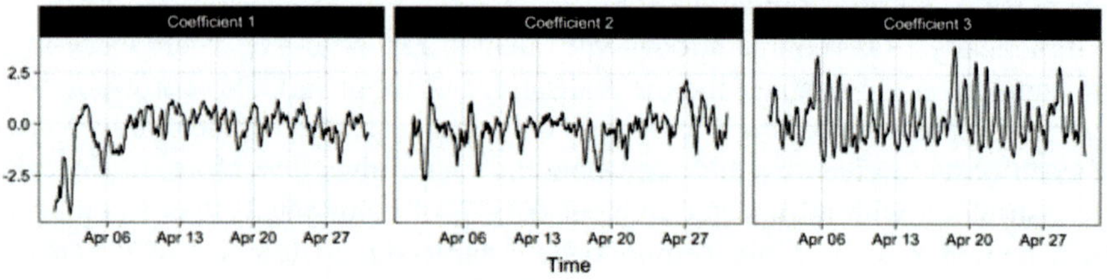

(C) The same coefficients zoom-in April 2015.

Fig. 3.1 **Spatial EOF decomposition**. The first three components of the EOF decomposition obtained from the MSWind 13–16 data set in the spatial formulation

3.1.2 Temporal Formulation

Although less common [8], the EOF analysis can also be performed analogously on the *empirical temporal covariance matrix*, from which temporal EOFs with *PC spatial coefficients* are obtained. This procedure is equivalent to performing a PCA with time indices considered as variables with spatial realisations. From this, spatio-temporal data can be represented in a linear combination of discrete temporal orthonormal basis elements where coefficients are purely spatial. More precisely, the *empirical temporal mean* at time t_j is defined as

$$\widehat{\mu}_t(t_j) := \frac{1}{S}\sum_{i=1}^{S} Z(\mathbf{s}_i, t_j). \tag{3.4}$$

For instance, Figures 2.10 and 2.11 displayed empirical temporal mean time series. The *temporally centred data*

$$\widetilde{Z}_t(\mathbf{s}_i, t_j) := Z(\mathbf{s}_i, t_j) - \widehat{\mu}_t(t_j), \tag{3.5}$$

can then be written as

$$\widetilde{Z}_t(\mathbf{s}_i, t_j) = \sum_k a_k(\mathbf{s}_i)\phi_k(t_j), \tag{3.6}$$

where the $\phi_k(t_j)$'s form a discrete temporal orthonormal basis and the $a_k(\mathbf{s}_i)$'s are the coefficients with respect to the kth temporal basis function ϕ_k at spatial locations $\mathbf{s}_i$, $i = 1, \dots, S$ such that

$$\begin{aligned}
&\mathbb{E}[a_k(\mathbf{s}_i)] = 0, \text{ for all } k, \\
&\mathrm{Var}\,[a_k(\mathbf{s}_i)] \geq \mathrm{Var}[a_{k+1}(\mathbf{s}_i)] \geq 0, \text{ for all } k, \\
&\mathrm{Cov}\,[a_k(\mathbf{s}_i),\ a_l(\mathbf{s}_i)] = 0, \text{ for all } k \neq l.
\end{aligned} \tag{3.7}$$

Here, the coefficients $a_k(\mathbf{s}_i)$'s are independent of time, while the temporal basis function $\phi_k(t_j)$ only depends on time and not on space location.

An EOF decomposition in the temporal formulation is performed on the same data set as the previous subsection and visualised in Fig. 3.2. It is observed that the temporal formulation does not provide the same decomposition as the spatial formulation. About 63% of the data variability is explained by the first three components. The first temporal EOF clearly shows yearly and daily cycles. It could be possible that the corresponding spatial coefficients, which are less structured than the second and third component coefficients, look like the third spatial EOF of Fig. 3.1a. It may suggest that the cycles are moderately spatially structured. The second and third temporal EOFs are harder to interpret, although one might see more variability in the winter. However, the spatial coefficients of the second component show an inversion of the behaviour of the second temporal EOF depending if the location is in the Alps or not. This spatial pattern is similar to the second spatial EOF of Fig. 3.1a. Finally, the spatial coefficients of the third component show a clear spatial structure.

3.1.3 Singular Value Decomposition

It is not advisable to compute the matrix product $\widetilde{\mathbf{Z}}^T\widetilde{\mathbf{Z}}$—and in particular the empirical covariance matrix of Eq. (3.2)—to get its eigenvalue decomposition. The Singular Value Decomposition (SVD) of $\widetilde{\mathbf{Z}}$ allows avoiding this, thanks to its relation to the spectral decomposition of $\widetilde{\mathbf{Z}}^T\widetilde{\mathbf{Z}}$—Eq. (3.3). This is discussed here in the context of EOF computation and will also appear in Chap. 6.

Consider the SVD of the rectangular real matrix $\tilde{\mathbf{Z}} \in \mathbb{R}^{T\times S}$,

$$\tilde{\mathbf{Z}} = \mathbf{U}^T\mathbf{D}\mathbf{V},$$

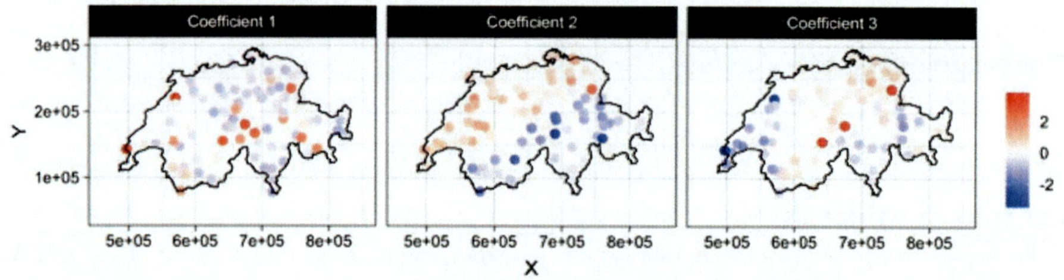

(A) Normalised PC spatial coefficients.

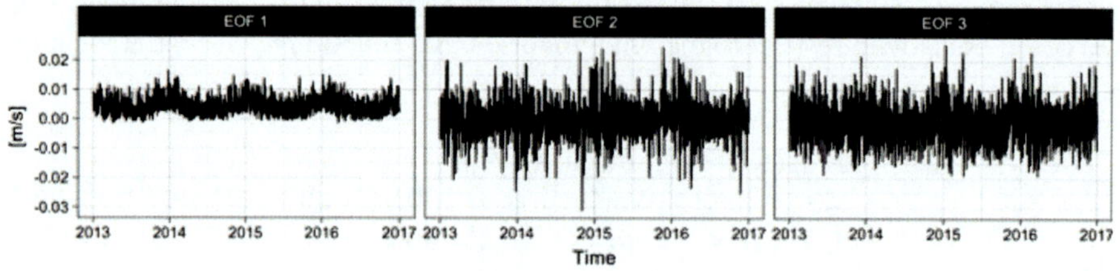

(B) Temporal EOF basis elements.

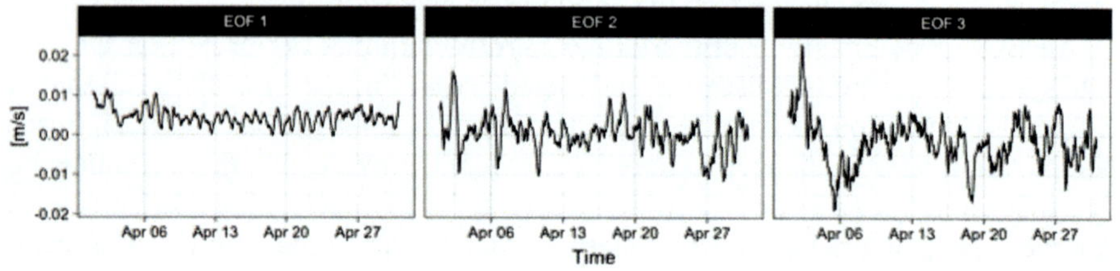

(C) The same basis elements zoom-in April 2015.

Fig. 3.2 Temporal EOF decomposition. The first three components of the EOF decomposition obtained from the MSWind 13–16 data set in the temporal formulation

with the orthogonal matrix of left singular vectors $\mathbf{U} \in \mathbb{R}^{T \times T}$, the rectangular matrix $\mathbf{D} \in \mathbb{R}^{T \times S}$ which contains the decreasing singular values down its main diagonal, and the orthogonal matrix of right singular vectors $\mathbf{V} \in \mathbb{R}^{S \times S}$. It is a well-known fact that the non-zero singular values of $\widetilde{\mathbf{Z}}$ are given by $\sqrt{\lambda_k}$, $k = 1, \ldots, S$ where λ_k are the non-zero eigenvalues of $\widetilde{\mathbf{Z}}^T\widetilde{\mathbf{Z}}$, and that the columns of $\mathbf{V}^T$ are eigenvectors of $\widetilde{\mathbf{Z}}^T\widetilde{\mathbf{Z}}$ [12].

In the particular case of EOFs computation, this means that $\mathbf{\Lambda} = \mathbf{D}^T\mathbf{D}$ and the columns of $\mathbf{V}^T$ are the EOFs—up to a sign. Additionally, the PC times series are obtained by

$$\mathbf{A} = \widetilde{\mathbf{Z}}\mathbf{V}^T = \mathbf{U}^T\mathbf{D}\mathbf{V}\mathbf{V}^T = \mathbf{U}^T\mathbf{D}.$$

From this, it is easy to see that the first S columns of the matrix $\mathbf{U}^T$ contain the normalised PC times series [8].

Finally, although EOFs can be computed from the spectral decomposition of the empirical spatial (or temporal) covariance matrix, an SVD of $\widetilde{\mathbf{Z}}$ is more efficient in practice [13] and allows to perform the decomposition even when $S > T$ (or $T > S$ in the temporal formulation). As a matter of fact, the SVD provides the matrix $\mathbf{V}^T$ which contains the EOFs, the matrix $\mathbf{D}$ which contains the square root of each component relative variance and yields directly the normalised temporal coefficients in matrix $\mathbf{U}^T$. See [8] for more details.

3.2 Variography

Although the semivariogram is the cornerstone of geostatistical modelling, it can also be used as an exploratory tool to describe the spatio-temporal correlation structures of a data set. Moreover, variography remains essential for analysing and interpreting the data, the model and the residuals of non-geostatistical approaches [14].

Consider the quantity

$$\gamma(\mathbf{s}_i, \mathbf{s}_k, t_j, t_l) := \frac{1}{2}\mathrm{Var}\left[Z(\mathbf{s}_k, t_l) - Z(\mathbf{s}_i, t_j)\right],$$

which is called the *spatio-temporal semivariogram*. The *intrinsic hypothesis* assumes that the spatio-temporal process under study has a constant expectation and that the spatio-temporal semivariogram depends only on the difference between spatio-temporal points, which is weaker than the stationary hypothesis [15]. In this case, one can write

$$\gamma(\mathbf{h}, \tau) := \frac{1}{2}\mathrm{Var}\left[Z(\mathbf{s}_i + \mathbf{h}, t_j + \tau) - Z(\mathbf{s}_i, t_j)\right],$$

with $\mathbf{h} = \mathbf{s}_k - \mathbf{s}_i$ is a spatial lag and $\tau = t_l - t_j$ is a temporal lag [4, 8]. If the spatio-temporal semivariogram depends only on the Euclidean distance $h = ||\mathbf{h}||$ but not on the direction of $\mathbf{h}$, it is said to be *isotropic*. If it is not isotropic, it is said to be *anisotropic*. The quantity $\lim_{\mathbf{h},\tau\to 0^+} \gamma(\mathbf{h}, \tau)$ is called the *nugget*. Under stationarity, the *sill* is defined as $\lim_{\mathbf{h},\tau\to\infty} \gamma(\mathbf{h}, \tau)$, while the *partial sill* is the difference between the sill and the nugget.

The *empirical spatio-temporal semivariogram* is given by

$$\widehat{\gamma}(\mathbf{h}, \tau) = \frac{1}{2 \cdot \#N_s(\mathbf{h}) \cdot \#N_t(\tau)} \sum_{\mathbf{s}_i,\mathbf{s}_k \in N_s(\mathbf{h})} \sum_{t_j,t_l \in N_t(\tau)} \left(Z(\mathbf{s}_k, t_l) - Z(\mathbf{s}_i, t_j)\right)^2,$$

where $N_s(\mathbf{h})$ is the set of all location pairs separated by a spatial lag of $\mathbf{h}$ within some tolerance, $N_t(\tau)$ is the set of all time points separated by a temporal lag of τ within some tolerance, and # denotes the cardinality of a set. It is the classical estimate of the spatio-temporal semivariogram [8, 15, 16]. All empirical spatio-

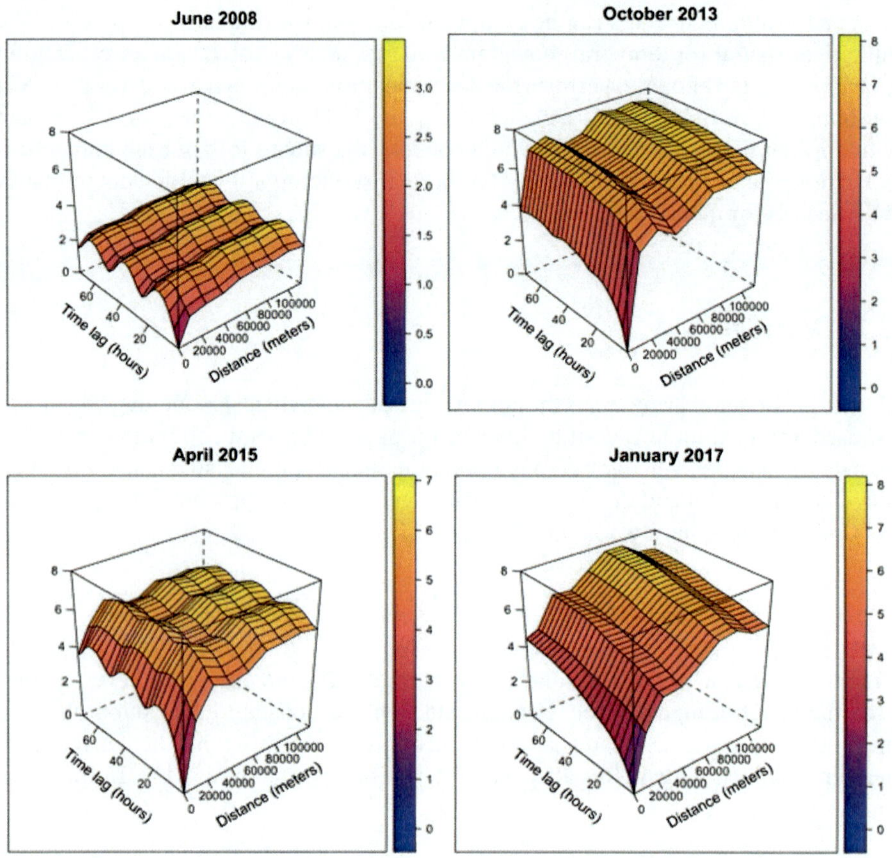

Fig. 3.3 Empirical spatio-temporal semivariograms for the MeteoSwiss wind speed data. For June 2008 (s^2 = 2.85), October 2013 (s^2 = 6.66), April 2015 (s^2 = 6.62) and January 2017 (s^2 = 6.61)

temporal semivariograms shown in this book are *omnidirectional*, that is, they are computed without considering direction (isotropy).

It is often helpful to compare the semivariogram with the *sample variance* of all the data, which will be noted s^2. Because of the correlation present in the data, this quantity should only be intended as a rough indication of the sill value [17]. When no spatio-temporal correlation is present, the empirical semivariogram should fluctuate around the sample variance up to small $\mathbf{h}$ and τ. This is the so-called *pure nugget effect*.

Figure 3.3 displays empirical spatio-temporal semivariograms for selected months of the MeteoSwiss wind speed data, numerically computed with the gstat R library [18, 19]. June 2008 has the lowest sill. The daily cycle is clearly marked along the time lag axis for June 2008 and April 2015. Its presence declines in October 2013 and completely disappears in January 2017. This is in good agreement with the corresponding time series visualised in Fig. 2.11. Spatial correlation is not striking for June 2008, October 2013 and April 2015, although for the latter, a spatial structure was highlighted with the EOF analysis of Sect. 3.1. However, a clear spatial structure is present for January 2017. Those variographic snapshots for different months showed that the spatio-temporal linear dependencies evolve with time.

A more detailed study performed by the thesis author with different data, but on the same phenomenon—wind speed in Switzerland from 2008 to 2017—also highlighted non-separability and anisotropy in the north-east direction likely corresponding to the characteristic channelling circulation of the wind in Switzerland [20]. Recall that non-separability indicates the impossibility of characterising the spatio-temporal covariance function by the product of purely temporal and purely spatial covariance functions. Non-separability was shown at daily, hourly, and 10 min frequencies with statistical hypotheses testing [21, 22] using the Covatest R package [23].

In Chaps. 5 and 7, empirical semivariograms will be used to understand the quality and the quantity of spatio-temporal dependencies extracted by the model from the original data through a residuals analysis [24].

3.3 Wavelet Variance Analysis

Wavelet variance is a signal processing method providing the decomposition of the variance of a signal by scales with the help of Multiresolution Wavelet Analysis (MRWA), which demonstrated to be a performing tool for extracting scale-dependent components of a time series [25]. This technique is applied to the MoTUS data set to discriminate the wind dynamics below and above the average building height around the mast. It will also allow an estimation of the wind speed power spectrum. Particularly for the higher anemometers, it will help to gain insight into wind speed in free field.

3.3.1 Multiresolution Wavelet Analysis

An accurate time series description in time and frequency is difficult with Fourier analysis. The wavelet transform overcomes this issue by decomposing the series on a basis generated by translating and scaling a function $\psi(t)$ called *mother wavelet*.

MRWA is a recursive procedure for performing discrete wavelet analysis [26]. Given a signal $x = \{x(i)\}$ of length T, sampled at regular intervals $\Delta t = \tau$, one can divide it into two components:

1. an approximated part A_1 of the signal x at the coarser scale $\Delta t = 2\tau$,
2. a detailed part D_1 at scale $\Delta t = \tau$.

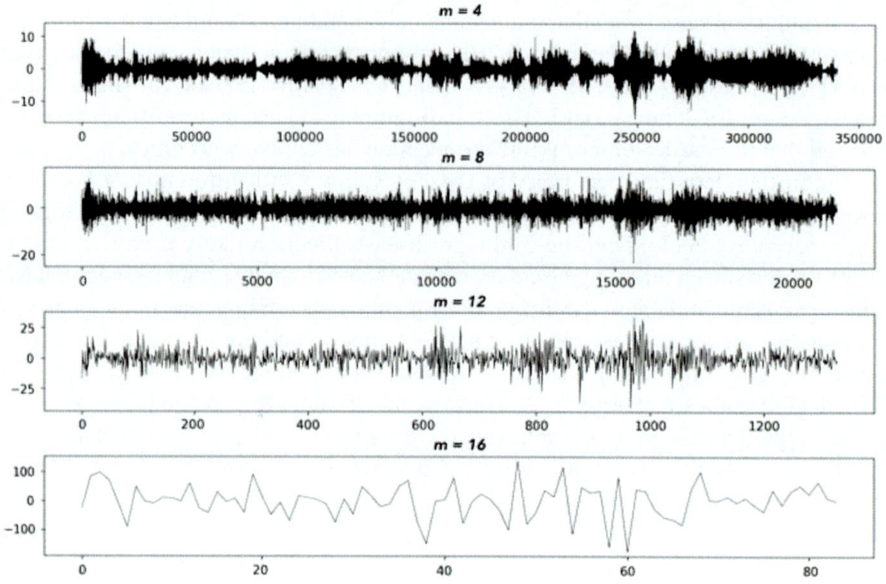

Fig. 3.4 Detailed coefficients of DWT. The scales $m = 4, 8, 12$ and 16 are visualised for the anemometer 1. Haar mother wavelet is used

This process is recursively applied on the approximated parts, each step providing an approximation A_m at scale $\Delta t = 2^m \tau$ and a detailed part D_m at scale $\Delta t = 2^{m-1}\tau$. After M iterations of the procedure, the time series is decomposed as

$$x = A_M + D_1 + D_2 + \cdots + D_M.$$

In practice, this whole iterative process is computed by performing the Discrete Wavelet Transform (DWT) algorithm [27]. Each detailed level D_m have $N_m = \text{int}(T/2^m)$ coefficients completely determined by the mother wavelet $\psi(t)$ and given by

$$d(m, n) = 2^{-m/2} \sum_{i=0}^{T-1} x(i)\psi(2^{-m}i - n),$$

for all $m = 1, \ldots, M$, $n = 0, \ldots, N_m - 1$ [28]. Detailed coefficients at different scale levels for Haar mother wavelet are shown in Fig. 3.4, as an example. Figure 3.5 displays a part of the 3D visualisation of the same coefficients. The coefficients at larger scales have a coarser resolution because the corresponding elements of the wavelet basis cover a more significant part of the original signal.

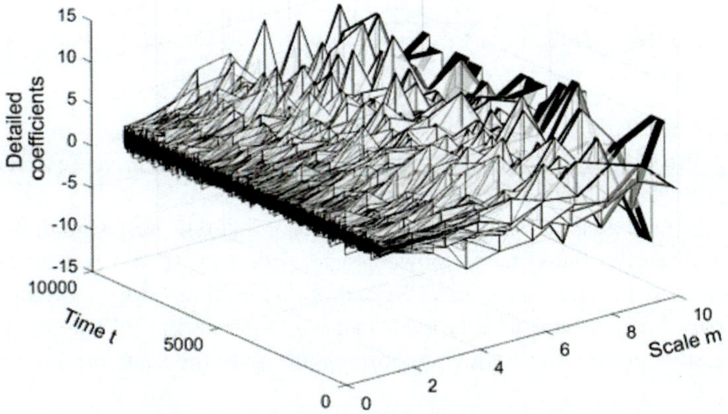

Fig. 3.5 Three-dimensional representation of the DWT detailed coefficients. The coefficients are presented as a function of scale over a part of the data for the anemometer 1, using Haar mother wavelet

3.3.2 The Wavelet Variance

The quantity of interest is the variance of the detailed coefficients for a given scale m, noted as $\sigma^2_{det}(m)$. An estimate for this quantity is defined by

$$\hat{\sigma}^2_{det}(m) = \frac{1}{N_m} \sum_{n=0}^{N_m-1} \left[d(m,n) - \frac{1}{N_m} \sum_{n=0}^{N_m-1} d(m,n) \right]^2, \tag{3.8}$$

for a given scale m. However, when $\hat{\sigma}_{det}(m)$ is normalised by 2^m, it provides a convenient scale decomposition of the time series variance. Following [29], let us define the *wavelet variance* as $\hat{\sigma}^2_{wav}(m) = \hat{\sigma}^2_{det}(m)/2^m$. Assume for convenience that $T = 2^M$. Supposing that the coefficient averages are zeros, then one has for all scale m,

$$\hat{\sigma}^2_{wav}(m) = \frac{\hat{\sigma}^2_{det}(m)}{2^m} = \frac{N_m}{T} \hat{\sigma}^2_{det}(m) = \frac{1}{T} \sum_{n=0}^{N_m-1} d(m,n)^2, \tag{3.9}$$

since $N_m = T/2^m$ and using Eq. (3.8). If the orthonormality of the basis induced by the mother wavelet is supposed, one can show [25, 29] that

$$\sum_{i=0}^{T-1} x(i)^2 = \sum_{m=1}^{M} \sum_{n=0}^{N_m-1} d(m,n)^2 + T\bar{x}^2, \tag{3.10}$$

where $\bar{x}$ denotes the average of $\{x(i)\}$. Hence, using Eqs. (3.9) and (3.10), one gets

$$\sum_{m=1}^{M} \hat{\sigma}^2_{wav}(m) = \frac{1}{T} \sum_{m=1}^{M} \sum_{n=0}^{N_m-1} d(m,n)^2 = \frac{1}{T} \sum_{i=0}^{T-1} x(i)^2 - \bar{x}^2. \tag{3.11}$$

Remark that the Right Hand Side (RHS) member of that equation is the sample variance of the time series. Hence, the detailed coefficients $d(m,n)$ of MRWA yield a natural scale decomposition of the variance by $\hat{\sigma}^2_{wav}(m)$.

In [30], Heneghan et al. show that for second-order stationary signal, the wavelet variance is directly related to the power spectral density (PSD) and give a simple interpretation of it: $\sigma^2_{wav}(m)$ is the PSD components filtered by a bandpass filter. The bandpass filter is determined by $\psi(t)$ and only passes the PSD components in a bandwidth surrounding the frequency corresponding to the scale m.

3.3.3 Application to the MoTUS Data

The wavelet variance is applied to the seven normalised wind speed series of the MoTUS data set. Four different mother wavelets are used: Haar, db2, db3 and db4. The wavelet variance for the seven time series is displayed in Fig. 3.6.

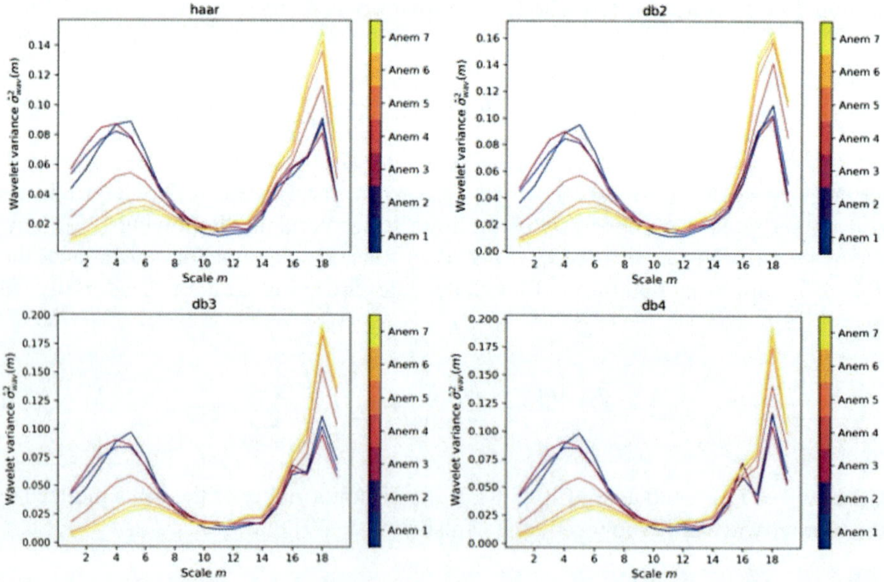

Fig. 3.6 Wavelet variance. Each panel displays wavelet variance for different mother wavelets applied to the normalised MoTUS time series

All wavelet variance plots show very similar characteristics:

1. The wavelet variance against scale m of the wind speed gathered by the three lower anemometers are almost overlapped, and curves show tiny differences between them;
2. Similarly, the wavelet variance against scale m of the wind speed gathered by the three higher anemometers are also overlapping, with very slight differences between each other;
3. The wavelet variance of the three lower anemometers is more significant at small scales, with a maximum between scales 4 and 5, while the wavelet variance of the three higher anemometers is bigger at large scales, with a maximum at the scale 18;
4. The wavelet variance curve corresponding to the wind speed measured at 13.5 [m] seems almost equidistant between the two groups of the three lower and the three higher wind measurement devices.

Interestingly, the wavelet variance curves of wind speed demonstrate its direct relation with wind speed PSD; see in particular Fig. 2.2 of [31] or Fig. 1 of [32]. The maxima around scale 5 are identifiable with microscale turbulences, while maxima around scale 18 correspond to wind speed variability associated with synoptic mean flows—with a period around 100 hours [31, 33]. Between them is the spectral gap, centred near 1/hour frequency—which is the frequency of the studied MeteoSwiss data.

From the perspective of atmospheric physics,[2] the three lower sensors capture intra-canyon dynamics, while the synoptic flow is most important above the urban canyon. Also, this may be related to the size of the turbulent eddies. The eddies are relatively small close to the ground [34, 35], while they increase linearly with height above the ground. These results suggest that the building layout impacts on the temporal variability of the wind: wavelet variance increases at smaller or larger scales for the lower or higher sensors, respectively. Considering the correspondence between wavelet scale and frequency—low/high frequency ranges are related to large/small wavelet scales—the lower anemometers are dominated by high-frequency fluctuations. This agrees with the typical small turbulence dynamics phenomena that occur around buildings close to the ground. Indeed, at small heights from the ground and in the proximity of buildings, the mechanical generation of turbulence is prevalent, and the airflow movements are dominated by small turbulent eddies [36, 37]. Located in an unconditioned flow state, the wind speed measured by the higher sensors is not characterised by intermittency or bursts of high-frequency fluctuations such as the wind measured at lower heights. Therefore, low-frequency fluctuations dominate its dynamics, which is more representative of the synoptic flows.

---

[2] The physical interpretations of the last paragraph of this subsection were provided by Dr. Dasaraden Mauree and Dr. Luciano Telesca.

3.4 Summary

This chapter presented some quantitative exploratory tools for spatio-temporal data, which partially confirmed the qualitative visual inspection done in Chap. 2.

EOF analysis unveiled patterns by projecting the spatio-temporal data onto a spatial (or temporal) basis. This kind of orthogonal basis decomposition will be the fundamental tool for the spatio-temporal modelling proposed in Chap. 5 of the book. Variography will be also used in Chaps. 5 and 7 to validate the model. Applied to the data as an EDA tool, variography analysis quantified linear dependencies in space and time. It revealed dynamical changes in the spatio-temporal dependencies, anisotropy, non-separability underpinning the complexity of the wind speed data. It also confirmed cyclicity when suspected by visual inspection of the time series. The wavelet variance analysis was applied to the MoTUS data set to better understand and quantify the urban zone's impact on wind speed behaviour. It provided discrimination between the lower and higher anemometers, which was unclear from basic EDAs of Chap. 2, through the ratio of turbulence and variability of synoptic flows.

The next chapter is entirely dedicated to methodological developments of a last advanced EDA method based on IT quantities, which will also contribute to exploring the wind speed data and gaining understanding from a distributional perspective.

References

1. Preisendorfer R (1988) Principal component analysis in meteorology and oceanog-raphy, Developments in atmospheric science. Elsevier
2. Hannachi A, Jolliffe I, Stephenson D (2007) Empirical orthogonal functions and related techniques in atmospheric science: a review. Int J Climatol: J R Meteorol Soc 27(9):1119–1152
3. Jolliffe IT, Cadima J (2016) Principal component analysis: a review and recent developments. Philos Trans R Soc A: Math, Phys Eng Sci 374(2065):20 150 202
4. Cressie N, Wikle C (2011) Statistics for spatio-temporal data. Wiley
5. Hristopulos DT (2020) Random fields for spatial data modeling. Springer
6. Sullivan TJ (2015) Introduction to uncertainty quantification, vol 63. Springer
7. Rasmussen CE, Williams CK (2006) Gaussian processes for machine learning. MIT press
8. Wikle C, Zammit-Mangion A, Cressie N (2019) Spatio-temporal Statistics with R, Chapman & Hall/CRC the R Series. CRC Press, Taylor & Francis Group
9. Bishop CM (2006) Pattern recognition and machine learning (Information science and statistics). Springer, Berlin, Heidelberg. isbn: 0387310738
10. Hsieh WW (2009) Machine learning methods in the environmental sciences: neural networks and kernels. Cambridge University Press
11. Jolliffe I (2013) Principal component analysis, Springer series in statistics. Springer New York. isbn: 9781475719048
12. Banerjee S, Roy A (2014) Linear algebra and matrix analysis for statistics. CRC Press
13. Gentle JE (2009) Computational statistics, vol 308. Springer
14. Kanevski M, Maignan M (2004) Analysis and modelling of spatial environmental data, vol 6501. EPFL Press
15. Montero J-M, Fernández-Avilés G, Mateu J (2015) Spatial and spatio-temporal geostatistical modeling and kriging, vol 998. Wiley

16. Sherman M (2011) Spatial statistics and spatio-temporal data: covariance functions and directional properties. Wiley
17. Chiles J-P, Delfiner P (2009) Geostatistics: modeling spatial uncertainty, vol 497. Wiley
18. Pebesma EJ (2004) Multivariable geostatistics in S: the gstat package. Comput Geosci 30:683–691
19. Gräler B, Pebesma E, Heuvelink G (2016) Spatio-temporal interpolation using gstat. R J 8(1):204–218
20. Guignard F, Kanevski M (2018) Spatio-temporal variography of wind speed in complex region. In: EGU general assembly conference abstracts, vol 20, p 4925
21. Li B, Genton MG, Sherman M (2007) A nonparametric assessment of properties of space-time covariance functions. J Am Stat Assoc 102(478):736–744
22. Cappello C, De Iaco S, Posa D (2018) Testing the type of non-separability and some classes of space-time covariance function models. Stoch Environ Res Risk Assess 32(1):17–35
23. De Iaco S, Cappello C, Posa P (2017) Covatest: tests on properties of space-time covariance functions. R package version 0.2. 1
24. Kanevski M, Pozdnoukhov A, Timonin V (2009) Machine learning for spatial environmental data. EPFL Press
25. Addison PS (2002) The illustrated wavelet transform handbook. Taylor & Francis
26. Gao J, Cao Y, Tung W-W, Hu J (2007) Multiscale analysis of complex time series. Wiley
27. Daubechies I (1992) Ten lectures on wavelets. Society for industrial and applied mathematics
28. Thurner S, Feurstein M, Teich MC (1997) Multiresolution wavelet analysis of heartbeat intervals discriminates healthy patients from those with cardiac pathology
29. Lindsay RW, Percival DB, Rothrock DA (1996) The discrete wavelet transform and the scale analysis of the surface properties of sea ice. Stoch Environ Res Risk Assess 34(3):771–787. issn: 0196-2892. https://doi.org/10.1109/36.499782
30. Heneghan C, Lowen SB, Teich MC (1999) Analysis of spectral and wavelet- based measures used to assess cardiac pathology. In: 1999 IEEE international conference on acoustics, speech, and signal processing. Proceedings. ICASSP99 (Cat. No. 99CH36258), vol 3, pp 1393–1396. https://doi.org/10.1109/ICASSP.1999.756241
31. Stull RB (1988) An introduction to boundary layer meteorology. Kluwer Academic Publishers, p 670
32. Van der Hoven I (1957) Power spectrum of horizontal wind speed in the frequency range from 0.0007 to 900 cycles per hour. J Atmos Sci 14(2):160–164
33. Whiteman C (2000) Mountain meteorology: fundamentals and applications. Oxford University Press
34. Santiago JL, Martilli A (2010) A dynamic urban canopy parameterization for mesoscale models based on computational fluid dynamics reynolds-averaged Navier-Stokes microscale simulations. Bound-Layer Meteorol 137:417–439. https://doi.org/10.1007/s10546-010-9538-4
35. Mauree D, Blond N, Kohler M, Clappier A (2016) On the coherence in the boundary layer: development of a canopy interface model. Front Earth Sci 4:109, issn: 2296-6463. https://doi.org/10.3389/feart.2016.00109
36. Coceal O, Belcher SE (2004) A canopy model of mean winds through urban areas. Q J R Meteorol Soc 130:1349–1372. https://doi.org/10.1256/qj.03.40
37. Xie Z-T, Coceal O, Castro IP (2008) Large-eddy simulation of flows over ran- dom urban-like obstacles. Bound-Layer Meteorol 129. https://doi.org/10.1007/s10546-008-9290-1

Chapter 4
Fisher-Shannon Analysis

This chapter discusses the Fisher-Shannon analysis, an effective and computationally efficient data exploration tool based on IT quantities. This method is highly versatile and can be used in various situations such as time series discrimination, complexity quantification, detection of non-linear relationships, generation of time series features, detection of dynamical changes and non-stationarity tracking of a signal. Section 4.1 reviews the origins of Fisher-Shannon analysis and its application to geosciences. Its concepts are presented in Sect. 4.2. Section 4.3 provides analytical formula in the particular cases of random variables following some positively skewed distributions, namely Gamma, Weibull and log-normal ones. Then, a non-parametric estimation of the related IT quantities is presented in Sect. 4.4. Experiments on both simulated and real-world data sets are performed in Sect. 4.5.

4.1 Related Work

Initially proposed for statistical estimation purposes [1], the FIM has been extensively used in theoretical physics by Frieden [2]. FIM and SEP [3] are closely related, as shown by IT [4, 5]. FSC—the FIM and SEP product—was proposed as a possible definition of atom complexity [6, 7].

Following Frieden work, FIM has found applications in non-linear time-series analysis. Martin et al. analysed complex non-stationary electroencephalographic signals and showed that FIM could have better discrimination performance than Shannon entropy [8]. FIM was also used to detect behaviour changes in dynamical systems [9]. Vignat and Bercher showed that a joint analysis of SEP and FIM could be required to discriminate non-stationary signals effectively [10].

F. Guignard, *On Spatio-Temporal Data Modelling and Uncertainty Quantification Using Machine Learning and Information Theory*, Springer Theses,
https://doi.org/10.1007/978-3-030-95231-0_4

The Fisher-Shannon method has been used to analyse complex dynamical processes in geophysics. Discrimination between the electric and magnetic components of magnetotelluric signals is performed in [11]. Tsunamigenic and non-tsunamigenic earthquakes were efficiently separated in the Fisher-Shannon information plane, using FSC [12]. Micro-tremors time series were identified depending on the soil characteristics of the measurement sites [13]. In [14], a classifier of (non-)tsunamigenic potential of earthquake built on several time series features—including FIM, SEP, FSC—is proposed. Finally, FIM was also used dynamically with sliding window techniques to study precursory patterns in seismology [15] and volcanology [16].

Many environmental processes have also been studied using the Fisher-Shannon method. In [17, 18], climatic regimes identification in rainfall time series is studied. Hydrological regimes discrimination has also been investigated [19]. Analysis of remotely sensed sea surface temperature showed in [20] that the Fisher-Shannon method can identify the Brazil-Malvinas Confluence Zone—which is known to be one of the most energetic areas of oceans. More than ten years of hourly wind speed data in the Fisher-Shannon information plane are analysed in [21]. The same authors studied the yearly variation of the FIM, the SEP and the FSC on wind measurements [22]. In [23], some pollutants are discriminated into the Fisher-Shannon plane, including cadmium, iron, and lead.

4.2 Fisher-Shannon Analysis

The fundamental research around the Fisher-Shannon method is somewhat scattered and comes from various fields, e.g. IT, physics, dynamical systems, ML and statistics. This section collects some known facts on FIM, SEP and their interactions. New interpretations are also proposed.

4.2.1 Shannon Entropy Power and Fisher Information Measure

Let us consider a univariate continuous random variable X with its Probability Density Function (PDF) $f(x)$, which is supposed to be sufficiently regular for the exposition of our purpose. Its *differential entropy* [5] is defined as

$$H_X = \mathbb{E}\big[-\log f(X)\big] = -\int f(x)\log f(x)\,dx. \tag{4.1}$$

For example, if X is a Gaussian random variable of variance σ^2, a direct computation gives $H_X = \frac{1}{2}\log(2\pi e\sigma^2)$. However, it will be more convenient to work with the following quantity, called the *Shannon Entropy Power* (SEP) [4],

$$N_X = \frac{1}{2\pi e} e^{2H_X}, \tag{4.2}$$

which is a strictly monotonically increasing transformation of H_X. The SEP is constructed such that $N_X = \sigma^2$ in the Gaussian case. Very often, entropies H_X and N_X are interpreted as global measures of disorder/uncertainty/spread of $f(x)$. The higher the entropy, the higher the uncertainty. The close connection between entropy and variance in the Gaussian case supports this interpretation [24].

The *Fisher Information Measure* (FIM) [10], also known as the *Fisher information of X with respect to a scalar translation parameter* [4], is defined as

$$I_X = \mathbb{E}\left[\left(\frac{\partial}{\partial x} \log f(X)\right)^2\right] = \int \frac{\left[\frac{\partial}{\partial x} f(x)\right]^2}{f(x)} dx. \tag{4.3}$$

This quantity should not be confused with the Fisher information of a distribution parameter. In particular, the derivative of the log-density is relative to x and not to some parameter. However, the FIM is equivalent to the Fisher information of a location parameter of a parametric distribution [5]. Under mild regularity conditions, one has the following alternative formulation [25],

$$I_X = \mathbb{E}\left[-\frac{\partial^2}{\partial x^2} \log f(X)\right]. \tag{4.4}$$

The quantity I_X is sometimes interpreted as a measure of order/organisation/narrowness of X. If X is Gaussian, $I_X = 1/\sigma^2$. The nomenclature suggests that the more informative a random variable, the less uncertain it is, as in the current language—and vice versa. Although the Gaussian case seems to confirm this interpretation, it could be more complex in other situations, as will be shown. It should be noted that H_X, N_X and I_X only depend on the distribution $f(x)$.

4.2.2 Properties

The SEP and the FIM respect several properties. First, both quantities are positive. It is also easy to see the *scaling properties* of the SEP and the FIM [26],

$$\begin{aligned} N_{aX} &= a^2 N_X, \\ I_{aX} &= a^{-2} I_X. \end{aligned} \tag{4.5}$$

for any real number $a \neq 0$, by a variable change. Notice also that the SEP and the FIM are invariant under additive deterministic constant, by the same argument. Harder to show are the *entropy power inequality* [4] and its dual the *Fisher information inequality* [27],

$$N_{X+Y} \geq N_X + N_Y, \tag{4.6}$$

$$I_{X+Y}^{-1} \geq I_X^{-1} + I_Y^{-1}, \tag{4.7}$$

for a random variable Y independent of X, with equality if X and Y are Gaussian.

Moreover, several relationships show that the FIM closely interacts with the SEP and the differential entropy. Let Z be a random variable independent of X with finite variance σ_Z^2. The *de Bruijn's identity* [5, 26] states that

$$\left. \frac{d}{dt} H_{X+\sqrt{t}Z} \right|_{t=0} = \frac{1}{2} \sigma_Z^2 I_X, \tag{4.8}$$

i.e. the variation of the differential entropy of a perturbed X is proportional to I_X. Therefore, a possible interpretation of the FIM is that it quantifies the sensitivity of H_X to a small independent additive perturbation Z [26]. Using the entropy power inequality (4.6) and de Bruijn identity (4.8), one can show the *isoperimetric inequality for entropies*,

$$N_X I_X \geq 1, \tag{4.9}$$

with equality if and only if X is Gaussian. The proof and the nomenclature motivation of Eq. (4.9) can be found in [4], where a remarkable analogy is made with geometry. These relationships show that SEP and FIM are intimately interlinked.

4.2.3 *Fisher-Shannon Complexity*

The joint FIM/SEP analysis has been used as a statistical complexity measure, albeit there is no clear consensus about the definition of signal complexity [7]. The *Fisher-Shannon Complexity* (FSC) is defined as [6]

$$C_X = N_X I_X. \tag{4.10}$$

From the scaling properties (4.5), it is easy to show that the FSC is constant under scalar multiplication and addition. In particular, normalisation or standardisation of X has no effect on the FSC. Additionally, the isoperimetric inequality for entropies (4.9) states that $C_X \geq 1$, with equality if and only if X is Gaussian. An interpretation of this quantity is the following. If Z is independent of X and has a finite variance σ_Z, one obtains the following relationship by using the de Bruijn identity (4.8),

$$\left. \frac{d}{dt} N_{X+\sqrt{t}Z} \right|_{t=0} = 2N_X \left. \frac{d}{dt} H_{X+\sqrt{t}Z} \right|_{t=0} = \sigma_Z^2 N_X I_X = \sigma_Z^2 C_X.$$

Hence, the FSC can be interpreted as a sensitivity measure of N_X to a small independent additive perturbation.

4.2.4 Fisher-Shannon Information Plane

The PDF of X can be analysed displaying the SEP and FIM within the so-called *Fisher-Shannon Information Plane* (FSIP), see Fig. 4.1 [10]. Although standard linear scale plots are very often used for the FSIP in the literature, a log-log plot is more adequate in practice. In the FSIP, the only reachable values are in the set $\mathcal{D} = \{(N_X, I_X) \in \mathbb{R}^2 | N_X > 0,\ I > 0 \text{ and } N_X I_X \geq 1\}$, due to (4.9). Reference [10] showed that for any point $(N, I) \in \mathcal{D}$, it exists a random variable X (from an exponential power distribution) such that $N_X = N$ and $I_X = I$.

A curve in $\mathcal{D}$ is said to be *iso-complex* if the FSC along the curve is constant. As C_X is constant up to a multiplicative factor $a \neq 0$, and looking up at the scaling properties (4.5), one can move on any iso-complex curve by varying a. Figure 4.1 shows the iso-complex curve of complexity $C_X = 10$ as an example. The boundary of $\mathcal{D}$ is the iso-complex curve with FSC equal to 1 and is reached if and only if X is Gaussian, as stated by (4.9). On this boundary, the standard deviation σ—which plays the role of the scaling parameter in the Gaussian case—is equivalent to the multiplicative factor a. Hence, while a point in the FSIP is described by (N_X, I_X), one can also describe it by (a, C_X). In the light of this, one can also think of FSC as a scale-independent measure of non-Gaussianity of X.

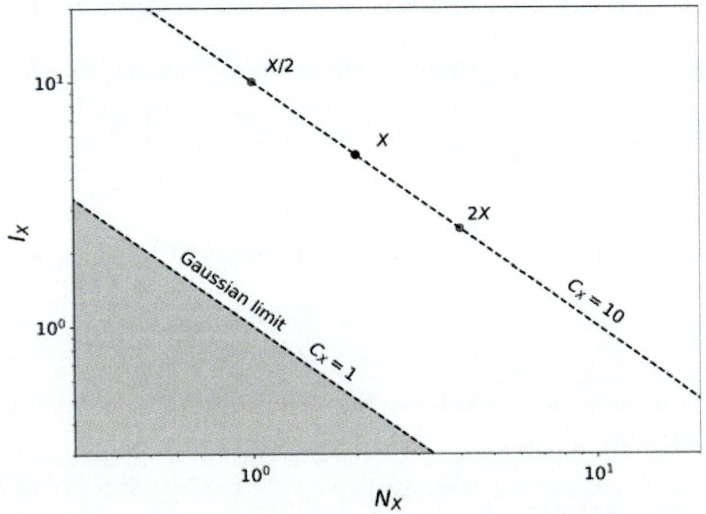

Fig. 4.1 The Fisher-Shannon information plane with a random variable X of FSC equal to 10. Scalar multiplication of X corresponds to a displacement along the iso-complex curve passing through X. The unreachable points are in grey. Note the logarithmic scale

4.3 Analytical Solutions for Some Distributions

In this section, analytical formulas for the SEP, FIM and FSC are proposed for several parametric distributions. They can be used directly for parametric estimations. Reference [10] obtained analogous formulas for the Student's t-distribution and the exponential power distribution (also known as generalised Gaussian distribution). The Gaussian case was already presented in Sect. 4.2 as an example.

The differential entropy of the distributions proposed in this section has been computed by [28], from which the SEP is directly obtained. However, to the author knowledge, the FIM-based calculations for Gamma, Weibull and log-normal distributions were never presented.

4.3.1 Gamma Distribution

The PDF of a Gamma random variable X is given by

$$f(x) = f(x; \theta, k) = \frac{x^{k-1} e^{-\frac{x}{\theta}}}{\theta^k \Gamma(k)}, \qquad \text{for } x \geq 0,$$

and $f(x) = 0$, for $x < 0$, where Γ denotes the Gamma function and $\theta, k > 0$ are respectively the scale and shape parameters.

The SEP of the Gamma distribution with scale $\theta > 0$ and shape $k > 0$ is

$$N_X(\theta, k) = \frac{\theta^2 \Gamma^2(k)}{2\pi e} e^{2[(1-k)\psi(k)+k]}, \tag{4.11}$$

where ψ is the digamma function.

The FIM is calculated for scale $\theta > 0$ and shape $k > 2$. Computing the derivative of $\log f(x)$, one has

$$\frac{\partial}{\partial x} \log f(x) = \frac{k-1}{x} - \frac{1}{\theta},$$

and then, by definition of FIM and using the variable change $x = \theta y$ and the Gamma function properties,

$$\begin{aligned}
I_X &= (k-1)^2 \mathbb{E}[X^{-2}] - \frac{2(k-1)}{\theta} \mathbb{E}[X^{-1}] + \frac{1}{\theta^2} \\
&= \frac{(k-1)^2}{\theta^k \Gamma(k)} \int_0^\infty x^{k-3} e^{-\frac{x}{\theta}} \, dx - \frac{2(k-1)}{\theta^{k+1} \Gamma(k)} \int_0^\infty x^{k-2} e^{-\frac{x}{\theta}} \, dx + \frac{1}{\theta^2} \\
&= \frac{(k-1)^2}{\theta^2 \Gamma(k)} \int_0^\infty y^{k-3} e^{-y} \, dy - \frac{2(k-1)}{\theta^2 \Gamma(k)} \int_0^\infty y^{k-2} e^{-y} \, dy + \frac{1}{\theta^2} \\
&= \frac{1}{\theta^2} \left[\frac{(k-1)\Gamma(k-2)}{(k-2)\Gamma(k-2)} - \frac{2\Gamma(k-1)}{\Gamma(k-1)} + 1 \right],
\end{aligned}$$

yielding

$$I_X(\theta, k) = \frac{1}{(k-2)\theta^2}. \tag{4.12}$$

An alternative derivation of the FIM can be obtained by computing the second derivative of $\log f(x)$,

$$\frac{\partial^2}{\partial x^2} \log f(x) = -\frac{k-1}{x^2},$$

and then, using (4.4) and the same variable change as before,

$$\begin{aligned} I_X &= (k-1)\mathbb{E}[X^{-2}] \\ &= \frac{k-1}{\theta^k \Gamma(k)} \int_0^\infty x^{k-3} e^{-\frac{x}{\theta}}\, dx \\ &= \frac{k-1}{\theta^2 \Gamma(k)} \int_0^\infty y^{k-3} e^{-y}\, dy \\ &= \frac{(k-1)\Gamma(k-2)}{\theta^2 (k-1)(k-2)\Gamma(k-2)}, \end{aligned}$$

yielding again equation (4.12).

The FSC of the Gamma distribution with scale $\theta > 0$ and shape $k > 2$ is directly obtained by multiplying Eqs. (4.11) and (4.12),

$$C_X(k) = \frac{\Gamma^2(k)}{2\pi e(k-2)} e^{2[(1-k)\psi(k)+k]}. \tag{4.13}$$

4.3.2 Weibull Distribution

The PDF of a Weibull random variable is

$$f(x) = f(x; \mu, \lambda, k) = \frac{k}{\lambda}\left(\frac{x-\mu}{\lambda}\right)^{k-1} e^{-(\frac{x-\mu}{\lambda})^k}, \qquad \text{for } x \geq 0,$$

and $f(x) = 0$, for $x < 0$, where μ is the location parameter, $\lambda > 0$ is the scale parameter and $k > 0$ is the shape parameter.

The SEP of the Weibull distribution with location μ, scale $\lambda > 0$ and shape $k > 0$ is

$$N_X(\lambda, k) = \frac{(1-\alpha)^2 \lambda^2 e}{2\pi} e^{2\alpha\gamma} \tag{4.14}$$

where $\alpha = \frac{k-1}{k}$ and γ is the Euler-Mascheroni constant.

The FIM is computed with location μ, scale $\lambda > 0$ and shape $k > 2$. The second derivative of $\log f(x)$ is

$$\frac{\partial^2}{\partial x^2} \log f(x) = -\frac{k-1}{(x-\mu)^2} - \frac{k(k-1)}{\lambda^k} (x-\mu)^{k-2},$$

and using Eq. (4.4) and the variable change $y = (\frac{x-\mu}{\lambda})^k$,

$$\begin{aligned}
I_X &= (k-1)\mathbb{E}\left[(X-\mu)^{-2}\right] + \frac{k(k-1)}{\lambda^k}\mathbb{E}\left[(X-\mu)^{k-2}\right] \\
&= \frac{k(k-1)}{\lambda^3}\left[\int_0^\infty \left(\frac{x-\mu}{\lambda}\right)^{k-3} e^{-(\frac{x-\mu}{\lambda})^k}\,dx + k\int_0^\infty \left(\frac{x-\mu}{\lambda}\right)^{2k-3} e^{-(\frac{x-\mu}{\lambda})^k}\,dx\right] \\
&= \frac{k-1}{\lambda^2}\left[\int_0^\infty y^{-\frac{2}{k}} e^{-y}\,dy + k\int_0^\infty y^{1-\frac{2}{k}} e^{-y}\,dy\right] \\
&= \frac{k-1}{\lambda^2}\left[\Gamma\left(1-\frac{2}{k}\right) + k\Gamma\left(2-\frac{2}{k}\right)\right] \\
&= \frac{k-1}{\lambda^2}\left[1 + k\left(1-\frac{2}{k}\right)\right]\Gamma\left(1-\frac{2}{k}\right) \\
&= \frac{(k-1)^2}{\lambda^2}\Gamma\left(1-\frac{2}{k}\right),
\end{aligned}$$

implying

$$I_X(\lambda, k) = \frac{\alpha^2}{(1-\alpha)^2\lambda^2}\Gamma(2\alpha - 1). \tag{4.15}$$

The FSC of the Weibull distribution with location μ, scale $\lambda > 0$ and shape $k > 2$ is then

$$C_X(k) = \frac{\alpha^2 e}{2\pi}\Gamma(2\alpha - 1)e^{2\alpha\gamma}. \tag{4.16}$$

4.3.3 Log-Normal Distribution

The log-normal PDF with parameters μ and $\sigma > 0$ is

$$f(x) = f(x; \mu, \sigma) = \frac{1}{x\sigma\sqrt{2\pi}} e^{-\frac{(\log x - \mu)^2}{2\sigma^2}}, \qquad \text{for } x > 0,$$

and $f(x) = 0$, for $x \leq 0$.

The notation of the parameters μ and σ are motivated by the fact that the logarithm of a log-normal random variable follows a normal distribution of mean μ and variance σ^2. However, μ and σ play respectively the role of the scale parameter and the shape parameter for the log-normal distribution.

The SEP is given by

$$N_X(\mu, \sigma) = \sigma^2 e^{2\mu}. \tag{4.17}$$

The second derivative of $\log f(x)$ is

$$\frac{\partial^2}{\partial x^2}\log f(x) = \frac{\log x - \mu + \sigma^2 - 1}{\sigma^2 x^2},$$

and using the variable change $y = \log x - \mu$, one has

$$\begin{aligned} I_X &= \frac{1}{\sigma\sqrt{2\pi}}\int_0^\infty \frac{1-\sigma^2-(\log x-\mu)}{\sigma^2 x^3} e^{-\frac{(\log x-\mu)^2}{2\sigma^2}}\,dx \\ &= \frac{1}{\sigma\sqrt{2\pi}}\int_{-\infty}^\infty \frac{1-\sigma^2-y}{\sigma^2} e^{-\frac{y^2}{2\sigma^2}-2y-2\mu}\,dy. \end{aligned}$$

Note that

$$-\frac{y^2}{2\sigma^2} - 2y - 2\mu = -\frac{(y+2\sigma^2)^2}{2\sigma^2} + 2(\sigma^2-\mu).$$

Using this and the definition of a Gaussian distribution $\mathcal{N}(-2\sigma^2, \sigma)$,

$$\begin{aligned} I_X &= \frac{1}{\sigma\sqrt{2\pi}}\int_{-\infty}^\infty \left(\frac{1-\sigma^2}{\sigma^2} - \frac{y}{\sigma^2}\right) e^{-\frac{(y+2\sigma^2)^2}{2\sigma^2}+2(\sigma^2-\mu)}\,dy \\ &= \frac{e^{2(\sigma^2-\mu)}}{\sigma^2}\left[\frac{1-\sigma^2}{\sigma\sqrt{2\pi}}\int_{-\infty}^\infty e^{-\frac{(y+2\sigma^2)^2}{2\sigma^2}}\,dy - \frac{1}{\sigma\sqrt{2\pi}}\int_{-\infty}^\infty y e^{-\frac{(y+2\sigma^2)^2}{2\sigma^2}}\,dy\right] \\ &= \frac{e^{2(\sigma^2-\mu)}}{\sigma^2}\left[1-\sigma^2+2\sigma^2\right] \\ &= \frac{1+\sigma^2}{\sigma^2} e^{2(\sigma^2-\mu)}, \end{aligned}$$

and the FIM is

$$I_X(\mu,\sigma) = \left(1+\frac{1}{\sigma^2}\right) e^{2(\sigma^2-\mu)}. \tag{4.18}$$

Multiplying Eq. (4.17) by Eq. (4.18), the FSC is then

$$C_X(\sigma) = (1+\sigma^2)e^{2\sigma^2}. \tag{4.19}$$

4.4 Data-Driven Non-parametric Estimation

Complex real-world data sets often do not follow parametric distributions. Providing enough data, it is also possible to carry out Fisher-Shannon analysis with a non-parametric estimation of density, which releases parametric assumptions on the distribution [29]. Here, *integral estimates* of the SEP and the FIM are considered, consisting of substituting the KDE of both $f(x)$ and its derivative in the integral forms of (4.1) and (4.3), [30–34]. Python and R implementations of this section content are proposed; see the software availability in Sect. 1.5.

Following [35], let $x_1, \ldots, x_n$ be a sample of size n from a PDF $f(x)$. Consider also the *kernel* $K(u)$, a bounded PDF which is symmetric around zero, has a finite fourth moment and is differentiable. The KDE of $f(x)$ is

$$\hat{f}_h(x) = \frac{1}{nh} \sum_{i=1}^{n} K\left(\frac{x - x_i}{h}\right), \tag{4.20}$$

where $h > 0$ is the *bandwidth* parameter. Here, the Gaussian kernel defined by $K(u) = (2\pi)^{-1/2} \exp(-u^2/2)$ is used and the estimator (4.20) becomes

$$\hat{f}_h(x) = \frac{1}{\sqrt{2\pi} nh} \sum_{i=1}^{n} \exp\left\{-\frac{1}{2}\left(\frac{x - x_i}{h}\right)^2\right\}.$$

The integral estimate of (4.2) is

$$\widehat{N}_X = \frac{1}{2\pi e} \exp\left\{-2 \int \hat{f}_h(x) \log \hat{f}_h(x)\, dx\right\}.$$

Let us note f', the derivative of f with respect to x. Usually, f' is estimated by $\hat{f}'_h$. With the Gaussian kernel, we obtain

$$\hat{f}'_h(x) = \frac{-1}{\sqrt{2\pi} nh^3} \sum_{i=1}^{n} (x - x_i) \exp\left\{-\frac{1}{2}\left(\frac{x - x_i}{h}\right)^2\right\}.$$

Then, the integral estimate of (4.3) is

$$\hat{I}_X = \int \frac{\left(\hat{f}'_h(x)\right)^2}{\hat{f}_h(x)}\, dx.$$

The FSC is estimated by multiplying $\widehat{N}_X$ by $\hat{I}_X$.

Several techniques exist to automatise the bandwidth choice [35]. In the following, the 2-stages direct plug-in method [36] is used. This method estimates the optimal bandwidth regarding the asymptotic mean integrated squared error of $\hat{f}_h$. The interested reader can find further technical details in [35, 36].

4.5 Experiments and Case Studies

In this section, various applications of SEP, FIM and FSC to time series are explored. First, a synthetic experiment is used to show the method's usefulness in detecting the dynamical behaviour of chaotic systems. The FSC of a multimodal combination of Gaussian distributions is investigated in a second experiment on a simulated data set. Then, examples of applications of the proposed method to real-world complex environmental data are discussed. Specifically, two detailed applications on the Motus data set are presented. Finally, three other contributions of the thesis author in the Fisher-Shannon analysis context are mentioned and briefly summarised.

4.5.1 Logistic Map

A synthetic experiment is designed to investigate how SEP, FIM and FSC can detect behavioural changes in nonlinear dynamical systems. In this section, the logistic map is used to illustrate and help understand important features of the considered measures.

Following the experiment proposed by [9], the *logistic map* defined by

$$x_{n+1} = cx_n(1 - x_n), \qquad x_0 \in [0, 1], \quad c \in [0, 4],$$

where c is the control parameter, is analysed using a sliding window technique. Analysis within the sliding window pursues the goal of revealing the dynamical evolution of time series properties like Gaussianity and non-stationarity.

The sequence $\{x_n\}$ is computed up to $n = 1000$ for $c \in [3.5, 4]$. Centred Gaussian noise with different level of variance, 0.05, 0.10, 0.15, is added to x_n. The well-known bifurcation diagram of the logistic map is displayed in Fig. 4.2. The SEP, FIM and FSC are computed on data included in the overlapping windows of width $2.5 \cdot 10^{-3}$ along with the control parameter, and the results are shown in the same Figure. The Lyapunov exponent is added for comparison reasons [37]. The results are also displayed in the FSIP; see Fig. 4.3.

The results obtained from the noiseless data analysis show how the SEP, FIM and FSC peak occurrences correspond to dynamic changes displayed by the bifurcation diagram and the Lyapunov exponent. With the logarithmic scale on the $y-$axis, the behaviour of the SEP is somewhat symmetric to the behaviour of the FIM, i.e. the FIM seems to be inversely proportional to the SEP. However, this is not exactly the case; otherwise, the FSC would be constant. In some sense, perturbations in the FSC reflect the departure from the inverse proportionality between SEP and FIM. In the FSIP, perfect inverse proportionality corresponds to iso-complex curves. Indeed, the trajectory of the logistic map in the FSIP is stretched along iso-complex curves; see Fig. 4.3.

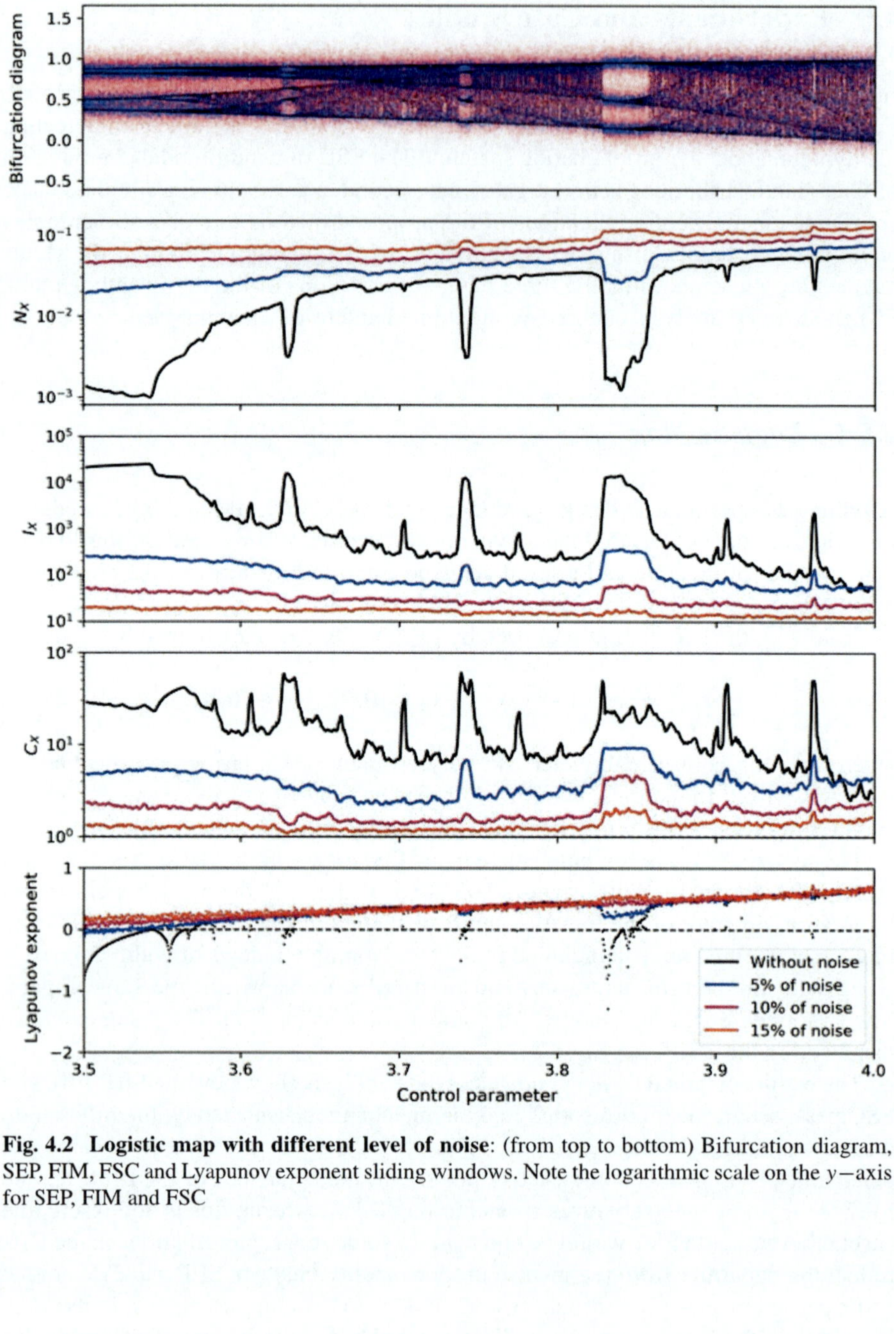

Fig. 4.2 Logistic map with different level of noise: (from top to bottom) Bifurcation diagram, SEP, FIM, FSC and Lyapunov exponent sliding windows. Note the logarithmic scale on the y−axis for SEP, FIM and FSC

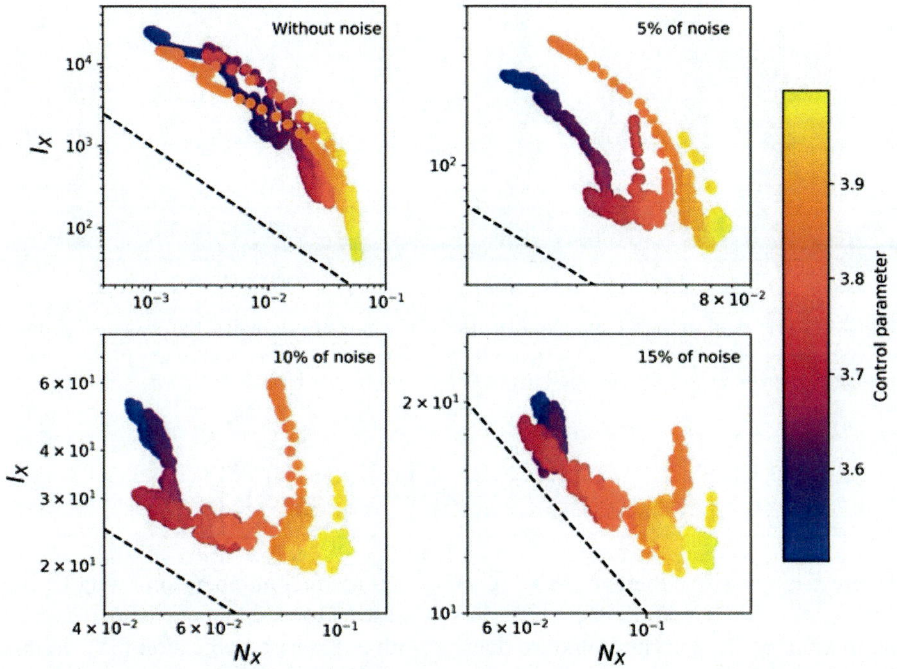

Fig. 4.3 Trajectory of the logistic map in the FSIP. Each panel displays a different amount of noise corruption. The Gaussian limit is in dashed line

Adding noise shows that most of the peaks become undetectable; see Fig. 4.2. However, FSC seems to be the measure that suffers the least to noise in data. Note also that FIM is less impacted than SEP. The noise effect is more interesting in the FSIP; see Fig. 4.3. While the uncorrupted data is quite hard to interpret due to the superposition of the trajectory with itself, adding some noise seems to clarify complexity and trajectory behaviours in the FSIP. Noise stimulates the emergence of protuberances roughly corresponding to "islands of stability" of the (uncorrupted) bifurcation diagram, where the Lyapunov exponent is negative. This emergence is due to the fact that the FIM is less impacted than the SEP, as seen above.

4.5.2 Normal Mixtures Densities

To better understand how multimodality affects the FSC, an original experiment on *normal mixture densities* is conducted. The density $f(x)$ belongs to such mixture if it writes as a convex combination of normal densities

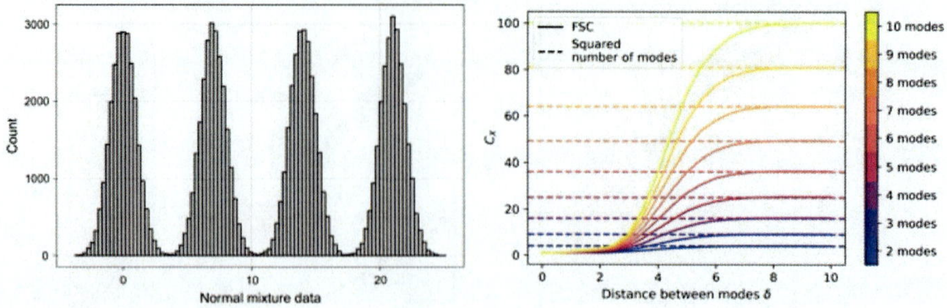

Fig. 4.4 Average of unitary Gaussian densities with equidistant modes: (left) An histogram of 100'000 points for $k = 4$ and $\delta = 7$. The resulting density has 4 modes and each pair of consecutive modes are separated by a distance of 7; (right) FSC for different k and varying δ

$$f(x) = \frac{1}{\sqrt{2\pi}} \sum_{l=1}^{k} \frac{w_l}{\sigma_l} \exp\left\{ -\frac{1}{2} \left(\frac{x - \mu_l}{\sigma_l} \right)^2 \right\},$$

where k is a positive integer, the weights $w_l \geq 0$ are real numbers summing to one and $\mu_l, \sigma_l^2 > 0$ are real numbers [38]. It is interesting to note that any density can be approximated by a normal mixture density with any desired precision [35]. In this specific experiment, the parameters are set to $w_l = 1/k$, $\sigma_l = 1$ and $\mu_l = (l - 1)\delta$, where δ represents a fixed distance between two consecutive modes, that is

$$f(x) = \frac{1}{\sqrt{2\pi}k} \sum_{l=1}^{k} \exp\left\{ -\frac{1}{2} (x - \mu_l)^2 \right\} = \frac{1}{k} \sum_{l=1}^{k} K(x - \mu_l). \tag{4.21}$$

In other words, $f(x)$ is an average of $\delta-$lagged unitary Gaussian densities. An example with $k = 4$ and $\delta = 7$ is displayed in the left panel of Fig. 4.4. Remark that in this particular setting, all modes have the same intensity.

The experiment studies the FSC of the density mentioned above when δ varies for different k. Given k and δ, $100'000$ points are generated following the density (4.21), from which the FSC is estimated. The right panel of Fig. 4.4 shows the results for $k = 2, 3 \ldots, 10$, and $\delta \in [0, 10]$. If $\delta = 0$, all modes overlay and the density corresponds to a Gaussian distribution with unitary FSC. If δ increases, the FSC grows during a somewhat transition phase before stabilising at a plateau of different magnitudes, depending on k. The end of the transition phase seems to correspond to sufficiently large δ such that each normal member of the mixture is no more overlapping significantly with its neighbours; see the histogram of the left panel, for example. Very surprisingly, the behaviour of the FSC when δ is increasing converges to k^2. In other words, if the mixture members are sufficiently separated, the FSC of $f(x)$ tends to the squared number of modes. In this situation, it suggests the existence of a relationship between the number of modes and the FSC; specifically, the number of modes would be $k = \sqrt{C_X}$.

Note that this empirical finding is not affected by a common scaling of the mixture members. The suggested relationship holds for an average of Gaussian density with the same variances and different distances between modes, provided that these distances are sufficiently large. Assuming that such a relationship exists in multivariate settings, this could open a way to study clustering problems, also known as unsupervised learning in the ML context.

4.5.3 *MoTUS Data: Advanced EDA*

Here, the Fisher-Shannon analysis is applied to explore the highest anemometer of the MoTUS data set. The Fisher-Shannon quantities are computed with non-overlapping moving windows of one-hour width along the time axis. Globally, all quantities vary with time, indicating non-stationarity; see Fig. 4.5. The SEP seems to replicate the behaviour of the original time series. A proportional effect between the mean and the variance of the data explains this behaviour, as shown in Fig. 4.6. As for the logistic map case, the FIM is roughly inversely proportional to the SEP (not shown in the logarithmic scale). The FSC is close to 1 during long periods, e.g. between the 17th and the 27th of January 2017. It should indicate a local behaviour of wind speed close to a Gaussian one. During these periods, wind speed is not necessarily calm, e.g. the 17th January. Conversely, The FSC also exhibits some peaks where wind speed is relatively low, which seems to indicate a more complex distribution of the data.

This intuition needs to be verified by a closer exploration of the data. To this aim, we considered four subsets of three-hour length, denoted by A, B, C and D and represented in Fig. 4.5 by the colours red, purple, blue and green, respectively. Histograms and Quantile-Quantile (Q-Q) plots of these data subsets are also plotted with the corresponding colours. Subset D is chosen during the period of almost unitary FSC. The corresponding histogram and Q-Q plot confirm the very-close-to-Gaussian behaviour of the data. Subset C is also chosen with an FSC close to 1 but centred on the SEP maximum of the 17th January 2017, also corresponding to a high wind speed activity. The histogram shows a distribution close to a Gaussian one again but with a higher variance than C. This output was expected since the SEP equals the variance for Gaussian distribution, and C was chosen with a high SEP. The Q-Q plot shows a slight departure for the left tail, but the data are still relatively close to what was expected. Subset B is centred on a peak of FSC. The histogram shows a distribution that is very far from Gaussianity. It is asymmetric and has at least two modes—maybe three. The Q-Q plot shows a substantial departure from the Gaussian distribution, especially on the left tail. Subset A is centred on the highest FSC value. Its histogram shows three—maybe four—modes. For this subset, the corresponding Q-Q plot shows how data are even farther from Gaussianity than the previous one.

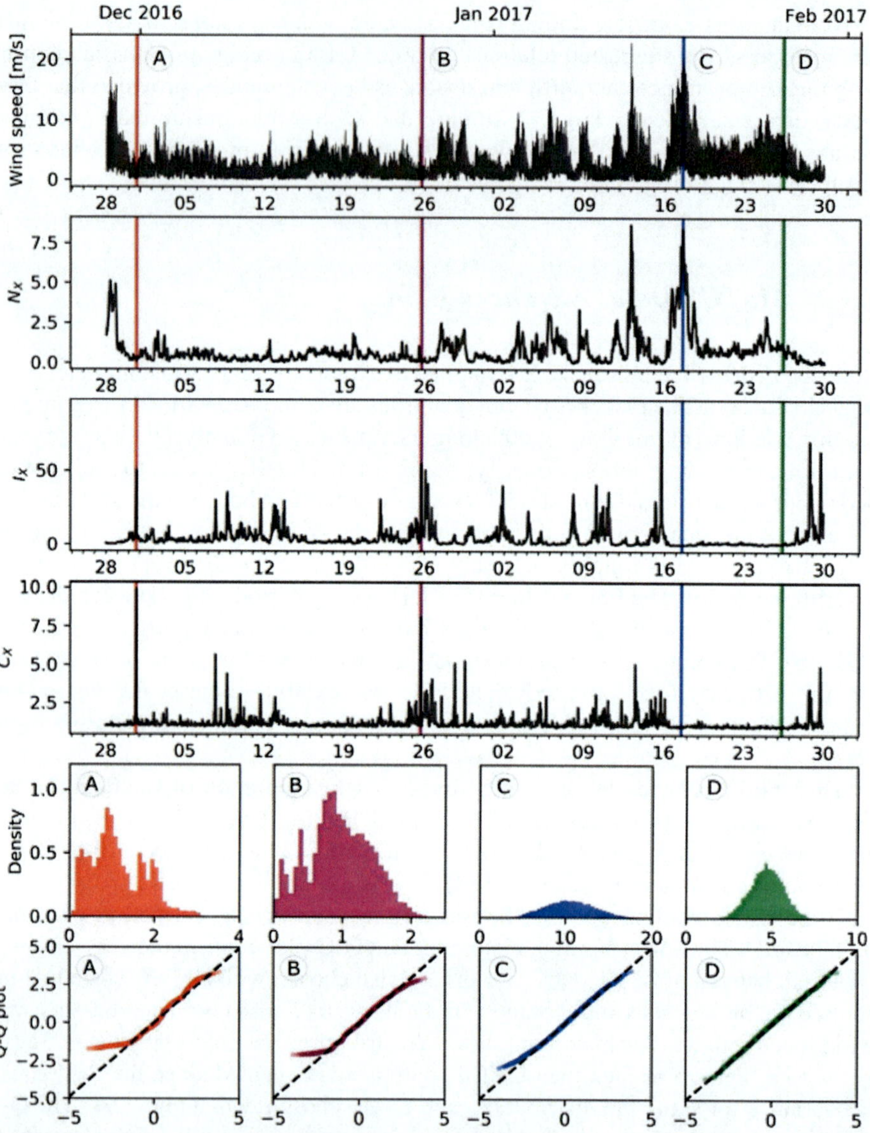

Fig. 4.5 Anemometer 7 of MoTUS dataset: (from top to bottom) High-frequency wind speed time series, SEP, FIM and FSC moving windows, histograms and Q-Q plots of some time series subsets. Note that FSC allows to identify Gaussian behaviour in the data

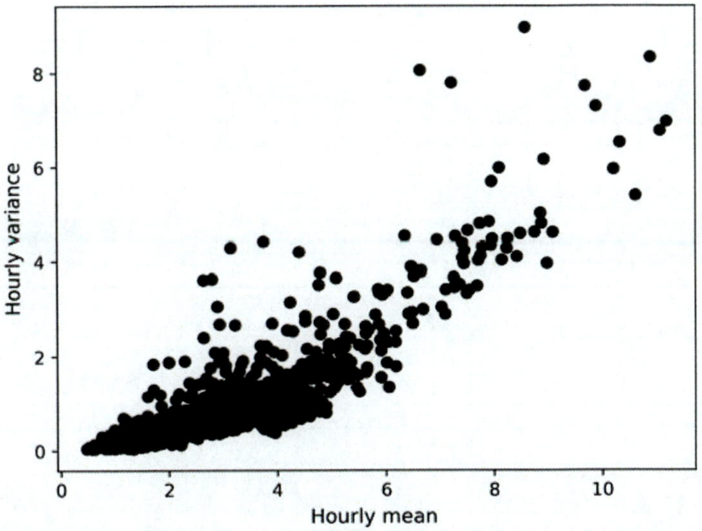

Fig. 4.6 Proportional effect. Scatter plot of hourly variance against hourly mean of the anemometer 7 of MoTUS data set

4.5.4 *MoTUS Data: Complexity Discrimination*

Further analysis of the whole MoTUS data set using the FSC can be found in this subsection. The FSC for each anemometer computed on the whole time series is shown in Fig. 4.7. There is a decreasing pattern of the FSC with height increasing

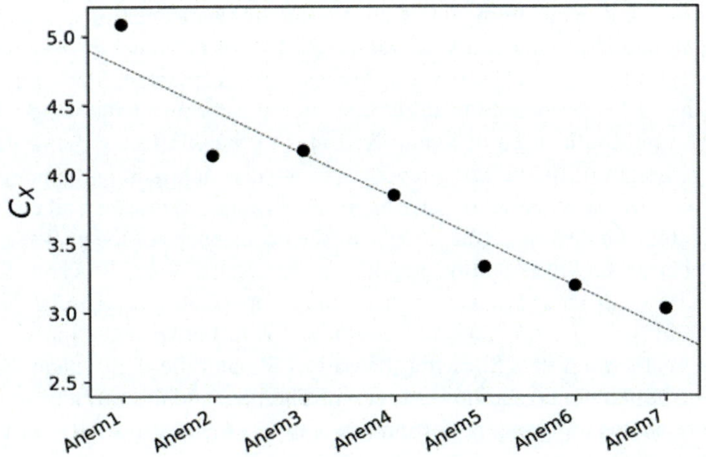

Fig. 4.7 FSC for all anemometers. Fisher-Shannon complexity of the 7 MoTUS wind speed time series

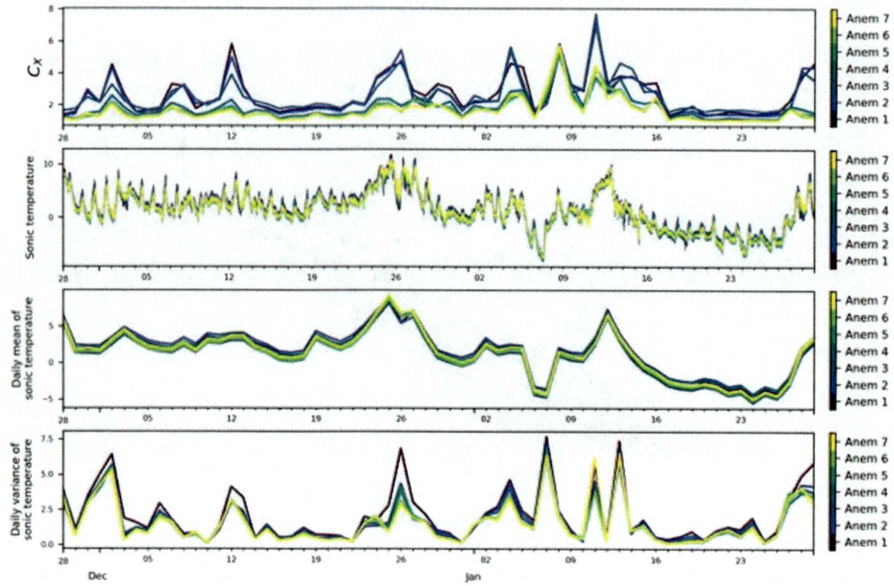

Fig. 4.8 Comparison with the sonic temperature: (from top to bottom) Daily FSC, sonic temperature in [°C], daily mean of sonic temperature and daily variance of sonic temperature at the 7 levels of the mast

from the ground. The largest value is reached at the lowest anemometer and the smallest for the highest one. In particular, this decrease demonstrates the existence of a non-linear relationship of wind speeds across height; otherwise, the FSC would be constant.

The FSC is also computed using the data of each day. The daily FSC—see the top of Fig. 4.8—reveals a clustering tendency among the seven anemometers: the lowest three anemometers are generally characterised by larger values of FSC during all the investigation period, while the four highest ones display smaller values of FSC through time. It is reasonable to think that such clustering of the wind series into two groups could reflect the different dynamics of the wind flow below and above the average height of the building layout[1] [39–42], which is in agreement with the findings based on the wavelet variance analysis of Chap. 3. However, all curves of the daily FSC somewhat show a coincidence of the occurrences of the peaks, especially in the last half of the investigation period.

The daily mean and variance of the sonic temperature are calculated and displayed in Fig. 4.8, to see if any link could be found between the daily FSC with weather/climatic variables. Showing the daily FSC and the daily mean sonic temperature, no apparent correlation is observed. The FSC and the variance of the sonic temperature shows an apparent correlation between them, especially for the lower

[1] In this subsection, all physical interpretations were provided by Dr. Dasaraden Mauree and Dr. Luciano Telesca.

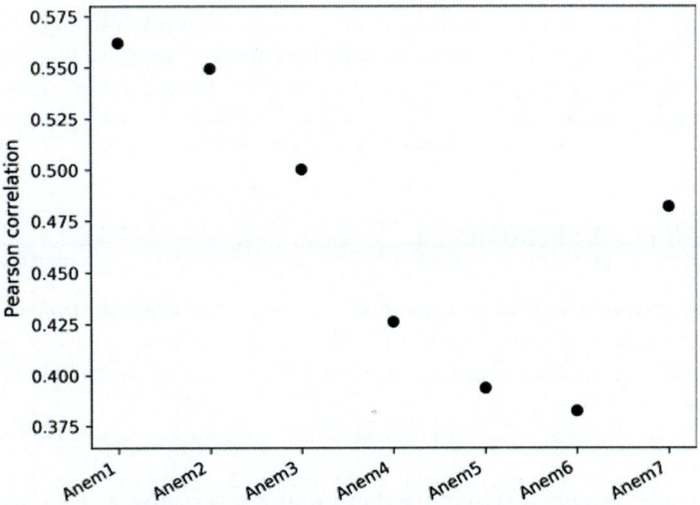

Fig. 4.9 Pearson correlation for all anemometers. Linear correlation between daily FS complexity and daily variance of sonic temperature at the 7 levels of the mast

Table 4.1 Statistical correlation test for all anemometers. Pearson correlation coefficient and p-value between daily FSC and daily variance of sonic temperature. The p-values are obtained with a non-parametric permutation test (999 permutations)

	Correlation	p-value
Anem 1	0.562	0.001
Anem 2	0.550	0.001
Anem 3	0.500	0.001
Anem 4	0.426	0.002
Anem 5	0.394	0.002
Anem 6	0.382	0.006
Anem 7	0.482	0.001

anemometers. Figure 4.9 shows that the Pearson correlation coefficient between the FSC and the variance of the sonic temperature is larger for the lower anemometers and smaller for the higher ones. Since the data are non-Gaussian, a non-parametric permutation test is performed for each anemometer in order to assess the significance of the correlation coefficients [43]. The number of permutations is set to 999. The results are presented in Table 4.1. It can be noted here that anemometer 7 presents a higher correlation. However, one possible explanation could be that anemometer 7 is in the inertial sub-layer, while 4–6 are in the roughness sub-layer and 1–3 are in the urban canopy layer. Nevertheless, further analysis should be done on more extended time series or at higher levels to help in understanding this particular phenomenon.

This analysis suggests that the temperature variation has a non-negligible effect on the wind speed's complexity—at least inside the canyon. The impact of the radiation and the differential heating of the surfaces inside the urban canyon could also lead to increased variances [42].

4.5.5 *Other Applications*

The Fisher-Shannon information method can find a wide application in the geo-environmental domains and beyond. This last subsection proposes to briefly summarise other results obtained through the Fisher-Shannon analysis to which the author of the thesis contributed.

4.5.5.1 Daily Means of Wind Speed in Complex Terrains

In [44], daily means of wind speed collected in Switzerland from 2012 to 2016 are investigated. The data are measured by 119 stations belonging to the MeteoSwiss SwissMetNet monitoring network and 174 stations belonging to the IMIS network, which covers the Alps densely and is managed by the WSL Institute for Snow and Avalanche Research SLF. The FIM and the SEP are computed, mapped over Switzerland and displayed in the FSIP. Relationships are identified between the IT quantities and both the elevation of the wind stations and the slope of the measuring sites. In particular, the results show that the FIM has its lowest values in the Alps, while the SEP has its highest values. Figure 4.10 visualises the FSIP where the colour changes with the terrain elevation.

4.5.5.2 Spatio-Temporal Evolution of Global Surface Temperature Distributions

In [45], it is proposed to use the Fisher-Shannon analysis to describe the non-stationarity of spatio-temporal climate fields. In particular, the complex behaviour of the maximum temperature at 2 [m] above the ground is highlighted. The NCEP CDAS1 daily reanalysis data from 1948 to 2018 covering the whole planet at a spatial resolution of $2.5 \times 2.5^{\circ}$ are used. For each spatial location, the temperature time series evolution is tracked with FIM, SEP and FSC through a sliding window of 5 years width and a sliding factor of 1 month, independently of its spatial neighbours. Each spatial location is represented dynamically in the FSIP—not shown here, see [45]. EOFs decomposition is also used to investigate the spatio-temporal fields of FIM and SEP, unveiling various spatial and temporal findings that may otherwise be difficult to recognise.

In particular, FSC is analysed in Fig. 4.11. The Hovmöller plot [46, 47] highlights how the equatorial areas exhibit an FSC fluctuating around values not far from 1,

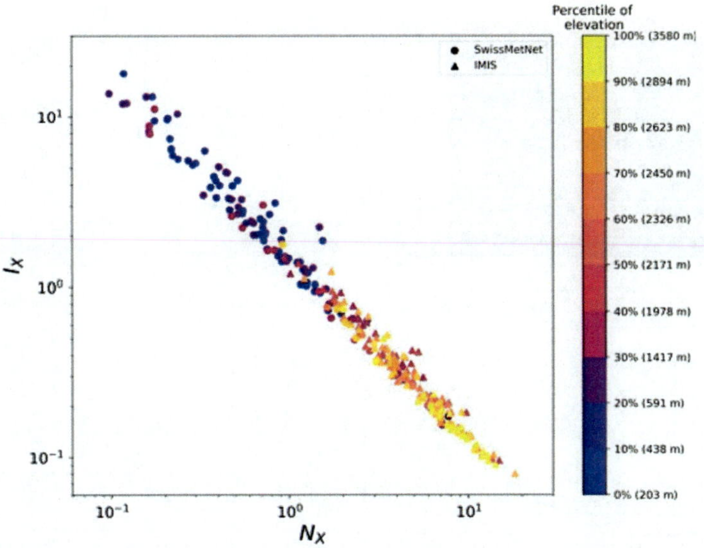

Fig. 4.10 FSIP of daily means of wind speed in Switzerland from 2012 to 2016. A non-linear dependence with altitude is clearly identifiable

implying temperature distributions almost Gaussian. Latitudes higher than 60° and lower than −60° show a reduction of the FSC over time. Figure 4.11 also investigates the FSC of the point having Longitude = 232.0 and Latitude = 56.2, located in western Canada. It is one of the points experiencing the most considerable overall change in the measured complexity over time among all the investigated spatial locations, obtained by summing up along time the squared differentiated time series and taking the maximum value among all pixels. This point is not far from Lytton, British Columbia, where the maximum temperature of the June-July 2021 Western North America heatwave was recorded. The distribution changes of temperature at this location are observed in the FSIP. The trajectory defined in the plane highlights different stages in its behaviour, which is also recognisable in the time series of the FSC. The latter shows decreasing complexity values since September 1989. As shown in the previous subsections, FSC can be used as an indicator of a changing pattern in data distribution. As an example, three distributions are shown, computed on the temporal windows from the 1st December 1951 to 30th November 1956, from the 1st September 1989 to the 31st August 1994 and from the 1st February 2009 to the 31st January 2014, respectively. All the distributions exhibit bimodality, where the seasonal variabilities determine the modes. The first distribution, associated with an FSC of 3.32, is not far from being a mixture of two Gaussian distributions, which is consistent with what we observed in Sect. 4.5.2. Differently, the second distribution plot, corresponding to an FSC of 29.37, is dominated by the mode at 273 [$°K$]. It is still valid in the last plot, corresponding to FSC of 6.67, although since September 1989, a reduction of the complexity has been registered, indicating a behaviour closer to Gaussian.

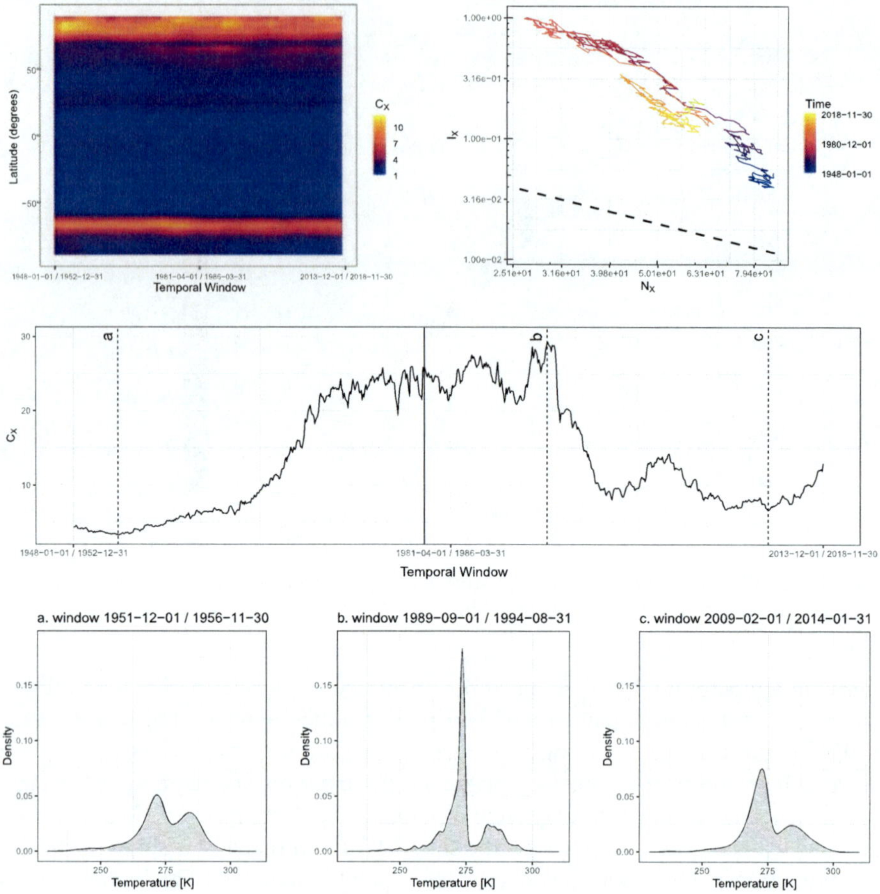

Fig. 4.11 **FSC of global surface temperature**: (top left) Hovmöller latitudinal plot; (top right) Trajectory in the FSIP of the point experiencing the greater overall change in the measured complexity over time having Longitude = 232.0 and Latitude = 56.2; (middle) FSC of the same point over time; (bottom row) Distributions of the temperature measurements in the temporal windows resulting into the FSC points highlighted with a vertical dashed line in the middle plot

4.5.5.3 Air Pollution Time Series Characterisation

In [45] hourly time series of three air pollutants are considered—Nitrogen dioxide (NO_2), Ground-level ozone (O3) and Particulate Matter (PM2.5). Those time series are collected at 16 monitoring stations in Switzerland. A relationship between the Fisher-Shannon analysis of three air pollutants PDF and measurement location in terms of land use and anthropogenic pollutant emission sources is emphasised. The findings are supported by clustering analysis, which identifies two different groups of time series, the one gathered in urban or the traffic-related zones and the one gathered in the rural areas. The statistical results are in line with the physical behaviour of the air pollution phenomenon.

4.6 Summary

After reviewing the literature of its fundamental aspect and its application to geosciences, the FIM, SEP and its joint use were presented. In particular, the close relationship between these two quantities was highlighted by equalities, inequalities and dualities in their properties. It led to the definition of FSC, which is used as a statistical complexity measure. All these quantities were conveniently represented in the FSIP. The discussed IT quantities were estimated by KDE regardless of the data distribution, although it could be evaluated by parametric estimation via the analytical formula proposed for several distributions.

Fisher-Shannon analyses were conducted on both simulated and real-world data sets. In particular, for the highest MoTUS anemometer, the use of sliding windows allowed to track its non-stationarity and detect local Gaussian behaviour. These results show the high complexity of these data, whose behaviour can rapidly change locally in time, even during calm weather. A deeper FSC analysis on the whole MoTUS data set suggested different wind dynamics induced by the building layout. A clear correlation between daily temperature variance and daily FSC of high-frequency wind speed records in urban area was found. Moreover, FSC was used to show that a non-linear relationship relates wind speed and height.

Numerous other applications have been presented, treating various data such as daily means of wind speed, air pollution time series and global surface temperature. For the latter, FSC was able to investigate both global and local distributional properties of spatio-temporal structures. Knowing that global warming is not uniform across the planet, such methods could be used to identify when and where patterns deviate from the average behaviour before investigating causal factors.

More generally, applications demonstrate the high versatility of analysis based on Fisher-Shannon information, yielding numerous and various insights on the data. While distributions are often studied by their moments, Fisher-Shannon analyse proposes a complementary perspective to traditional EDA tools.

References

1. Fisher RA (1925) Theory of statistical estimation. Math Proc Cambridge Philos Soc 22(5):700–725. https://doi.org/10.1017/S0305004100009580
2. Frieden BR (1990) Fisher information, disorder, and the equilibrium distributions of physics. Phys Rev A 41:4265–4276, 8 Apr 1990. https://doi.org/10.1103/PhysRevA.41.4265
3. Shannon CE (1948) A mathematical theory of communication. Bell Syst Tech J 27(3):379–423 (1948). https://doi.org/10.1002/j.1538-7305.1948.tb01338.x

4. Dembo A, Cover TM, Thomas JA (1991) Information theoretic inequalities. IEEE Trans Inf Theory 37(6):1501–1518. issn: 0018-9448. https://doi.org/10.1109/18.104312
5. Cover TM, Thomas JA (2006) Elements of information theory (Wiley series in telecommunications and signal processing). Wiley- Interscience, New York, NY, USA, isbn: 0471241954
6. Angulo J, Antolín J, Sen K (2008) Fisher-Shannon plane and statistical complexity of atoms. Phys Lett A 372(5):670–674, issn: 0375-9601. https://doi.org/10.1016/j.physleta.2007.07.077
7. Esquivel RO, Angulo JC, Antolín J, Dehesa JS, López-Rosa S, Flores-Gallegos N (2010) Analysis of complexity measures and information planes of selected molecules in position and momentum spaces. Phys Chem Chem Phys 12(26):7108–7116. https://doi.org/10.1039/B927055H
8. Martin M, Pennini F, Plastino A (1999) Fisher's information and the analysis of complex signals. Phys Lett A 256(2):173–180. issn: 0375-9601. https://doi.org/10.1016/S0375-9601(99)00211-X
9. Martin M, Perez J, Plastino A (2001) Fisher information and nonlinear dynamics. Phys A: Stat Mech Its Appl 291(1):523–532, issn: 0378-4371. https://doi.org/10.1016/S0378-4371(00)00531-8
10. Vignat C, Bercher J-F (2003) Analysis of signals in the Fisher-Shannon information plane. Phys Lett A 312(1):27–33. issn: 0375-9601. https://doi.org/10.1016/S0375-9601(03)00570-X
11. Telesca J-F, Lovallo M, Hsu H-L, Chen C-C, Analysis of dynamics in magnetotelluric data by using the Fisher-Shannon method. Phys A: Stat Mech Its Appl 390(7):1350–1355, issn: 0378-4371. https://doi.org/10.1016/j.physa.2010.12.005
12. Telesca L, Lovallo L, Chamoli L, Dimri L, Srivastava K (2013) Fisher-Shannon analysis of seismograms of tsunamigenic and non-tsunamigenic earth- quakes. Phys A: Stat Mech Its Appl 392(16):3424–3429. issn: 0378-4371. https://doi.org/10.1016/j.physa.2013.03.049
13. Telesca L, Lovallo L, Alcaz L, Ilies I (2015) Site-dependent organization structure of seismic microtremors. Phys A: Stat Mech Its Appl 421:541–547, issn: 0378-4371. https://doi.org/10.1016/j.physa.2014.11.061
14. Telesca L, Chamoli A, Lovallo M, Stabile TA (2015) Investigating the tsunamigenic potential of earthquakes from analysis of the informational and multifractal properties of seismograms. Pure Appl Geophys 172(7):1933–1943, issn: 1420-9136. https://doi.org/10.1007/s00024-014-0862-3
15. Telesca L, Lovallo M, Ramirez-Rojas A, Angulo-Brown F (2009) A non- linear strategy to reveal seismic precursory signatures in earthquake-related self-potential signals. Phys A: Stat Mech Its Appl 388(10):2036–2040, issn: 0378-4371. https://doi.org/10.1016/j.physa.2009.01.035
16. Telesca L, Lovallo M, Carniel R (2010) Time-dependent Fisher information measure of volcanic tremor before the 5 April 2003 paroxysm at stromboli Volcano, Italy. J Volcanol Geotherm Res 195(1):78–82, issn: 0377-0273. https://doi.org/10.1016/j.jvolgeores.2010.06.010
17. Lovallo M, Shaban A, Darwich T, Telesca L (2013) Investigating the time dynamics of monthly rainfall time series observed in northern lebanon by means of the detrended fluctuation analysis and the Fisher-Shannon method. Acta Geophys 61(6):1538–1555, issn: 1895-7455. https://doi.org/10.2478/s11600-012-0094-9
18. Pierini JO, Scian B, Lovallo M, Telesca L (2011) Discriminating climato- logical regimes in rainfall time series by using the Fisher-Shannon method. Int J Phys Sci 6(34):7799–7804
19. Pierini JO, Restrepo JC, Lovallo M, Telesca L (2015) Discriminating between different streamflow regimes by using the Fisher-Shannon method: an application to the Colombia rivers. Acta Geophys 63(2):533–546. issn: 1895-7455. https://doi.org/10.2478/s11600-014-0229-2
20. Pierini JO, Lovallo M, Gomez EA, Telesca L (2016) Fisher-Shannon analysis of the time variability of remotely sensed sea surface temperature at the brazil-malvinas confluence. Oceanologia 58(3):187–195, issn: 0078-3234. https://doi.org/10.1016/j.oceano.2016.02.003
21. Telesca L, Lovallo M (2011) Analysis of the time dynamics in wind records by means of multifractal detrended fluctuation analysis and the Fisher-Shannon information plane. J Stat Mech: Theory Exp 2011(07):P07001. https://doi.org/10.1088/1742-5468/2011/07/p07001

22. Telesca L, Lovallo M (2013) Fisher-Shannon analysis of wind records. Int J Energy Stat vol 01(04):281–290
23. Telesca L, Caggiano R, Lapenna V, Lovallo M, Trippetta S, Macchiato M (2009) Analysis of dynamics in Cd, Fe, and Pb in particulate matter by using the Fisher-Shannon method. Water, Air, Soil Pollut 201:33–41
24. Le ND, Zidek JV (2006) Statistical analysis of environmental space-time processes. Springer Science & Business Media
25. Lehmann E (1999) Elements of large-sample theory, Springer Texts in Statistics. Springer, New York, isbn: 978-0-387-98595-4
26. Rioul O (2011) Information theoretic proofs of entropy power inequalities. IEEE Trans Inf Theory 57(1):33–55, issn: 0018-9448. https://doi.org/10.1109/TIT.2010.2090193
27. Zamir R (1998) A proof of the Fisher information inequality via a data processing argument. IEEE Trans Inf Theory 44:1246–1250
28. Lazo AV, Rathie P (1978) On the entropy of continuous probability distributions (corresp.). IEEE Trans Inf Theory 24(1):120–122, issn: 0018-9448. https://doi.org/10.1109/TIT.1978.1055832
29. Telesca L, Lovallo M (2017) On the performance of Fisher information measure and Shannon entropy estimators. Phys A: Stat Mech Its Appl 484:569–576, issn: 0378-4371. https://doi.org/10.1016/j.physa.2017.04.184
30. P. K. Bhattacharya PK (1967) Estimation of a probability density function and its derivatives. Sankhyā: Indian J Stat, Ser A (1961–2002), 29(4):373–382
31. Dmitriev Y, Tarasenko F (1973) On the estimation of functionals of the probability density and its derivatives. Theory Probab Its Appl 18(3):628–633. https://doi.org/10.1137/1118083
32. Prakasa Rao B (1983) Nonparametric functional estimation, Probability and mathematical statistics: a series of monographs and textbooks. Academic Press
33. Györfi L, van der Meulen EC (1987) Density-free convergence properties of various estimators of entropy. Comput Stat Data Anal 5(4):425–436, issn: 0167-9473. https://doi.org/10.1016/0167-9473(87)90065-X
34. Joe H (1989) Estimation of entropy and other functionals of a multivariate density. Ann Inst Stat Math 41(4):683–697. issn: 1572-9052. https://doi.org/10.1007/BF00057735
35. Wand M, Jones M (1994) Kernel smoothing. Chapman & Hall/CRC Monographs on Statistics & Applied Probability. Taylor & Francis, isbn: 9780412552700
36. Sheather SJ, Jones MC (1991) A reliable data-based bandwidth selection method for kernel density estimation. J R Stat Soc Ser B (Methodological) 53(3):683–690, issn: 00359246
37. Kantz H, Schreiber T (2004) Nonlinear time series analysis. Cambridge university Press, vol 7
38. Marron JS, Wand MP (1992) Exact mean integrated squared error. Ann Stat 712–736
39. Coceal O, Belcher SE (2004) A canopy model of mean winds through urban areas. Q J R Meteorol Soc 130:1349–1372. https://doi.org/10.1256/qj.03.40
40. Christen A, van Gorsel E, Vogt R (2007) Coherent structures in urban roughness sublayer turbulence. Int J Climatol 27:1955–1968. https://doi.org/10.1002/joc.1625
41. Mauree D, Blond N, Kohler M, Clappier A (2017) On the coherence in the boundary layer: development of a canopy interface model. Front Earth Sci 4:109, issn: 2296-6463. https://doi.org/10.3389/feart.2016.00109
42. Oke TR, Mills G, Christen A, Voogt JA (2017) Urban climates. Cambridge University Press
43. Davison AC, Hinkley DV (1997) Bootstrap methods and their application, Cambridge series in statistical and probabilistic mathematics. Cambridge University Press. https://doi.org/10.1017/CBO9780511802843
44. Guignard F, Lovallo M, Laib M et al (2019) Investigating the time dynamics of wind speed in complex terrains by using the Fisher-Shannon method. Phys A 523:611–621
45. Amato F, Guignard F, Humphrey V, Kanevski M (2020) Spatio-temporal evolution of global surface temperature distributions. In: Proceedings of the 10th international conference on climate informatics, pp 37–43. https://doi.org/10.1145/3429309.3429315
46. Hovmöller E (1949) The trough-and-ridge diagram. Tellus 1(2):62–66
47. Wikle CK, Zammit-Mangion A, Cressie N (2019) Spatio-temporal Statistics with R, Chapman & Hall/CRC the R series. CRC Press, Taylor & Francis Group

Chapter 5
Spatio-Temporal Prediction with Machine Learning

This chapter proposes a methodological framework for spatio-temporal interpolation that enables the use of any ML algorithms through data decomposition into a convenient temporal basis with spatial coefficients. For example, a basis induced by EOFs is chosen, and a multiple-output deep Feed-forward Neural Network (FNN) is used to learn the spatial coefficients jointly. Across several different experimental settings using both simulated and real-world data, it is shown that the proposed framework allows reconstructing of coherent spatio-temporal fields. The chapter is structured as follows. In Sect. 5.1, motivation is further specified. In Sect. 5.2, the framework for the decomposition of spatio-temporal data and the modelling of the obtained random spatial coefficients is introduced. The proposed methodology is then applied to synthetic data in Sect. 5.3, temperature data in Sect. 5.4 and wind speed data in Sect. 5.5. Each application is discussed and validated through the analysis of its residuals.

5.1 Motivation

Environmental spatio-temporal data are usually characterised by spatial, temporal, and spatio-temporal correlations, as for wind speed; see previous chapters. Capturing these dependencies is extremely important.

Provided a relevant, well-designed feature space, ML algorithms—including ANNs—have proven that they can overcome this challenge for purely spatial data [1]. Deep Learning (DL) is of particular interest, primarily because of its capability to automatically extract features [2]. Most of the efforts have been focused on transposing methodologies from computer vision to the study of climate or environmental raster data—i.e. measurements of continuous or discrete spatio-temporal fields at regularly fixed locations [3]. These data are generally coming from satellites for earth observation or from climate models outputs. Nevertheless, especially at the

F. Guignard, *On Spatio-Temporal Data Modelling and Uncertainty Quantification Using Machine Learning and Information Theory* , Springer Theses,
https://doi.org/10.1007/978-3-030-95231-0_5

local or regional scales, there could be several reasons to prefer to work with direct ground measurements—i.e. measurements of continuous spatio-temporal fields on a set of irregular points in space—rather than with raster data [4]. Indeed, ground data are more precise than satellite products and climate models output, generally with a higher temporal sampling frequency of phenomena and without missing data due to e.g. clouds. On the other hand, the direct adaptation of models coming from traditional ML is more complex in the case of spatio-temporal data collected over sparsely distributed measurement stations. In this case, the approaches discussed in the literature to date only permit to perform modelling at the locations of the measurement stations and—to the best of the author knowledge—no ANN-based methodology has been proposed to solve spatio-temporal interpolation problems having as input spatially irregular ground measurements.

In order to fill this gap, the present methodological framework proposes to reconstruct a spatio-temporal field on a regular grid using spatially irregularly distributed time series. The spatio-temporal process of interest is described in temporally referenced basis functions with corresponding spatially distributed coefficients. The latter are considered stochastic, and the spatial coefficients' estimation is reformulated in terms of a set of regression problems based on spatial covariates.

5.2 Machine-Learning-Based Framework for Spatio-Temporal Interpolation

When working with spatio-temporal phenomena, it is difficult to realistically reproduce the spatial, temporal and spatio-temporal dependencies present in the data. One way of keeping into account these dependencies is to adopt a basis function representation [5]. In the following, details about the spatio-temporal signal decomposition and the structure of the regression problem to be solved to perform the spatial coefficient interpolation are discussed.

5.2.1 Decomposition of Spatio-Temporal Data Using EOFs

Assume that the temporally centred data $\widetilde{Z}_t(\mathbf{s}_i, t_j)$ defined in Eq. (3.5) are represented in a discrete temporal basis with purely spatial coefficients

$$\widetilde{Z}_t(\mathbf{s}_i, t_j) = \sum_k a_k(\mathbf{s}_i)\phi_k(t_j).$$

The EOF temporal formulation that was already encountered in Sect. 3.1, Chap. 3, provides such decomposition. However, many other temporal basis functions could be used, such as Fourier, Wavelets, Gaussian functions. Basis functions can be classified by different properties such as orthogonality, global/local, multi-resolutional behaviour, or data-driven ones. EOFs are bi-orthogonal basis functions. In particular, the covariance between coefficients vanishes, as specified by Eqs. (3.7), while the basis functions are orthogonal. Another attractive property of EOFs is its optimality regarding MSE of truncated reconstruction of the random field with the first K components of the basis [6]. As already mentioned in Chap. 3 and stated in Eqs (3.7), the data variance explained by each basis element decreases at each component. Basis truncation reduces the dimensionality of the model [5, 7].

In these regards, the spatio-temporal data are assumed to follow

$$Z(\mathbf{s}_i, t_j) = \widehat{\mu}_t(t_j) + \sum_{k=1}^{K} a_k(\mathbf{s}_i)\phi_k(t_j) + \eta(\mathbf{s}_i, t_j), \tag{5.1}$$

where $\widehat{\mu}_t(t_j)$ is the empirical temporal mean defined by Eq. (3.4), $\phi_k(t_j)$ are the first K EOFs basis coming from the temporally centred data, $a_k(\mathbf{s}_i)$ are the corresponding coefficients, and $\eta(\mathbf{s}_i, t_j)$ is a centred error term which contains any random part not captured by the model and possibly includes spatial and temporal dependencies.

While in ML PCA—or equivalently EOFs—is generally applied to reduce the dimensionality of the input space, here it is used on the output space—i.e. the spatio-temporal target variable that needs to be interpolated—to decompose the data into fixed temporal bases and their corresponding spatial coefficients. These latter are then considered as the target variable in a spatial regression problem, which can, in principle, be solved using any ML technique.

5.2.2 *Spatial Modelling of the Coefficients*

The—potentially truncated—EOFs decomposition returns for each spatial location $\mathbf{s}_i$, corresponding to the original observations, K random coefficients $a_k(\mathbf{s}_i)$. These coefficients can be spatially modelled and mapped on a regular grid solving an interpolation/regression task.

The coefficients will be modelled using a fully connected deep FNN [2] to demonstrate the effectiveness of the proposed approach. The structure of the network is described in Fig. 5.1. Spatial covariates are used as inputs for the neural network having a first auxiliary output layer where the spatial coefficients are modelled. A recomposition layer will then use the K modelled coefficients and the temporal bases ϕ_k resulting from the EOFs decomposition in order to reconstruct the final output—i.e. the spatio-temporal field—following Eq. (5.1), and noted $\widehat{Z}(\mathbf{s}_i, t_j)$.

The described network has multiple inputs, namely the spatial covariates—which flow through the whole stack of layers—and the temporal bases directly connected to the output layer. It also has multiple outputs: the spatial coefficients for each basis, all modelled jointly, and the output signal. While the network is trained by minimising

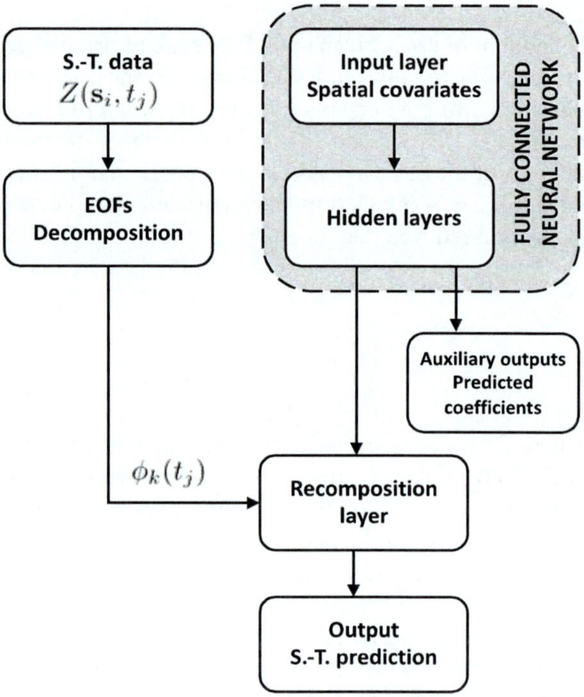

Fig. 5.1 Architecture of the proposed framework. The temporal bases are extracted from a decomposition of the spatio-temporal signal using EOFs. Then, a fully connected neural network is used to learn the corresponding spatial coefficients. The whole spatio-temporal field is recomposed following Eq. (5.1) and used for loss minimisation

the loss function on the final output, having the spatial coefficients maps as auxiliary output ensures a better explainability of the model.

It has to be highlighted that instead of this specific proposed DL structure,[1] any other traditional ML algorithm could have been used to model the spatial coefficients as a standard regression problem, indicating rather interesting flexibility of the proposed framework. Specifically, the framework will be used with a particular type of single-layer FNN, which enable UQ—namely ELM—in Chap. 7. However, the DL approach has two main advantages. The first is that most classical ML regression algorithms cannot handle multiple outputs, and thus one would have to fit separate models for each coefficient map without taking advantage of the similarities between the tasks. The second advantage is that the proposed DL approach minimises the loss directly on the final prediction target, i.e. the reconstructed spatio-temporal field of interest. Differently, when using single-output models, each is separately trained to minimise a loss computed on the spatial coefficients. It should, in principle, leads to a lower performance than with the proposed DL approach minimising directly the error computed on the full reconstructed spatio-temporal signal.

The following experiments are conducted with a fully connected 6-layers FNN, which has been implemented in Tensorflow [8]. Each layer contains 100 neurons

[1] The idea of using a DL structure was originally proposed by Dr. Federico Amato, who implemented and performed the prediction calculations.

with the ELU activation function. The ANN is regularised using an early stopping procedure. Other optimisation and implementation details can be found in [9].

5.3 Simulated Data Case Study

To produce 2-dimensional spatial patterns, 20 Gaussian random fields with Gaussian kernel are simulated from the RandomFields R package [10] and noted $X_k(\mathbf{s})$, $k = 1, \ldots, 20$. Time series $Y_k(t_j)$ of length $T = 1080$, for $k = 1, \ldots, 20$, are generated using an order 1 autoregressive model. Then, the simulated spatio-temporal data set is obtained as a linear combination of the spatial random fields $X_k(\mathbf{s})$, where $Y_k(t_j)$ play the role of the coefficients at time t_j, i.e.

$$Z(\mathbf{s}, t_j) = \sum_{k=1}^{20} X_k(\mathbf{s}) Y_k(t_j) + \varepsilon, \tag{5.2}$$

where ε is an independent and identically distributed (i.i.d.) noise term generated from a Gaussian distribution having zero mean and standard deviation equal to the 10% of the standard deviation of the noise-free field. Spatial points $\mathbf{s}_i$, $i = 1, \ldots, S = 2000$ are sampled uniformly on a regular 2-dimensional spatial grid of size 139×88, which will constitute the training locations. The spatio-temporal training set $\{Z(\mathbf{s}_i, t_j)\}$ is generated by evaluating the sequence of fields (5.2) at the training locations. Spatio-temporal validation and testing sets are generated analogously from 1000 randomly selected locations each.

The training and validation sets are decomposed following the methodology described in the previous section. The cumulative percentage of the relative variance for the first 50 components is represented in Fig. 5.2. Note that without the addition of the ε term, the total variance would have been explained with the first 20

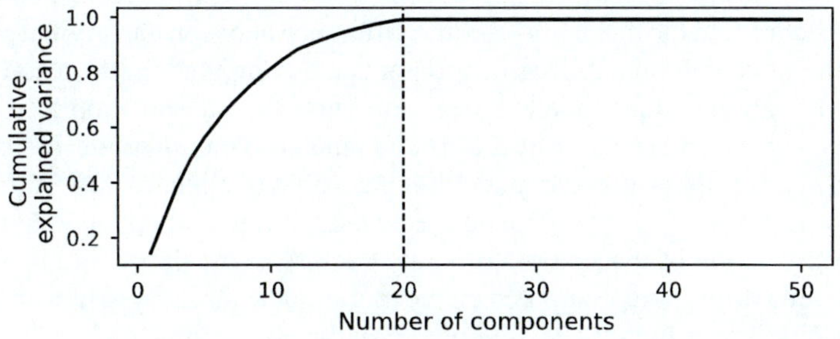

Fig. 5.2 Cumulative percentage of explained variance for the simulated data set. The variance explained by the first 50 components of the EOFs decomposition is displayed. As expected from the data generating process, the sum of the relative variances of the 20 first components reaches the total variance of the data

Table 5.1 MAE test on simulated and temperature data sets. MAE computed on the testing set, resulting from a model following a multiple-output strategy—in which all the spatial coefficients are learned at once—and the one obtained through an approach in which each coefficient map is predicted using a separate single-output regression model

Data set	Multiple outputs	Single output
Simulated	**1.978**	8.340
Temperature (all comp.)	**1.148**	1.599
Temperature (24 comp.)	**1.285**	1.628

components, which is the number of elements used to construct the simulated data set; see Eq. (5.2). Nonetheless, with the additional noise, these components explain about 99% of the variance, consistently with the 10% of the noise-free field standard deviation. Therefore, the neural network is trained using two inputs, namely the spatial covariates corresponding to the x and y coordinates and $K = 20$ outputs. The loss function is minimised regarding the Mean Absolute Error (MAE).

The MAE computed on the testing set after the modelling is indicated in Table 5.1. The output of the proposed framework is also compared to the results obtained using a different strategy. Instead of predicting all spatial coefficients at once with the proposed multiple-outputs model, the impact of doing separate modelling of each coefficient map on testing error performance is investigated. To this aim, fully connected FNN has been used to predict the individual spatial coefficient maps, which are then used together with the temporal bases to reconstruct the spatio-temporal field. The neural network has the same structure as the one used with the multiple-output strategy, except for the recomposition layer, which is indeed absent. It is shown that the use of the proposed multiple-output model helps to significantly improve the performances with respect to the approaches based on separate single-output models.

Randomly chosen examples of prediction maps and time series are shown and compared with the true spatio-temporal field in Fig. 5.3. The predicted map recovers the true spatial pattern, and the temporal behaviours are fairly well replicated too.

Figure 5.4 shows the spatio-temporal semivariograms of the simulated data, the model prediction and the residuals—i.e. the difference between the original simulated data and the modelled one. All semivariograms are computed on the testing points. The semivariogram on the modelled data shows how the interpolation recovered the same spatio-temporal structure of the true simulated data, although its values are slightly lower. It implies that the model has been able to explain most of the spatio-temporal variability of the phenomenon. However, it must be pointed out that even better reconstruction of the spatio-temporal structure of the data could be recognisable in the semivariograms computed on the training set, similarly to how the training error would be lower than the testing one. Finally, almost no structure is shown in the residuals semivariogram, suggesting that almost all the spatially and temporally structured information—or at least the one described by the semivariogram—has been extracted from the data. It also shows a nugget corresponding to the noise used in the generation of the data set.

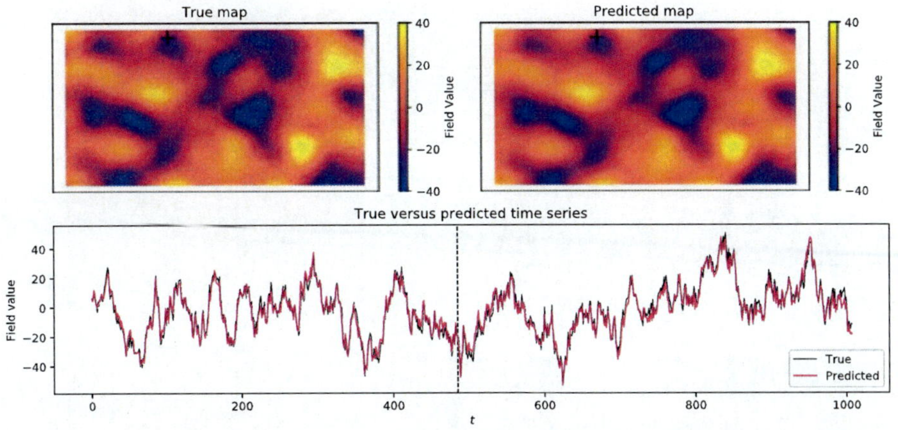

Fig. 5.3 Model prediction for the simulated spatio-temporal field defined by Eq. (5.2): (top left) A snapshot of the true spatial field at the fixed time indicated by the vertical dashed line in the temporal plot below; (top right) The predicted map at the same time; (bottom) The true time series (in black) and the predicted time series (in magenta) at the fixed location marked by a cross in the maps above

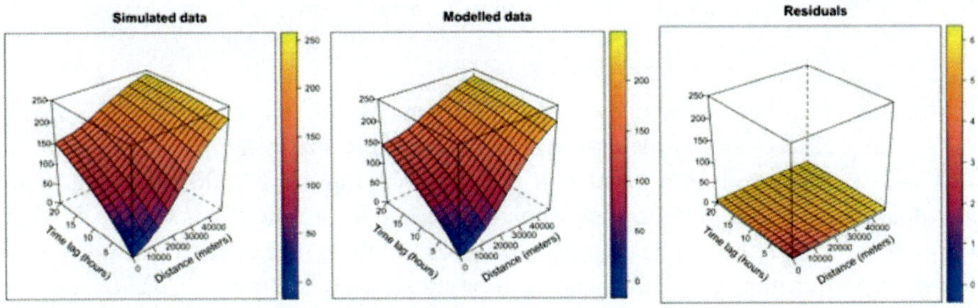

Fig. 5.4 Variography for the simulated data set. Spatio-temporal semivariograms have been computed on the testing points for the simulated data : (left) Original data, s^2 = 245; (centre) Prediction modelled using the first 20 EOFs components, s^2 = 237; (right) Residuals, s^2 = 6.4

5.4 Experiment on Temperature Monitoring Network

The effectiveness of the proposed framework in modelling real-world environmental phenomena is tested on a case study of air temperature prediction in Switzerland. Temperature is a crucial parameter in Earth system monitoring and modelling. Although less complicated to model than wind speed, air temperature can still be a challenging example, as it depends on elevation, topography, has daily and yearly cycles and can change abruptly because of air mass advection [11]. Moreover, the barrier formed by the Alps leads to marked differences in temperature between the two sides of the mountain range.

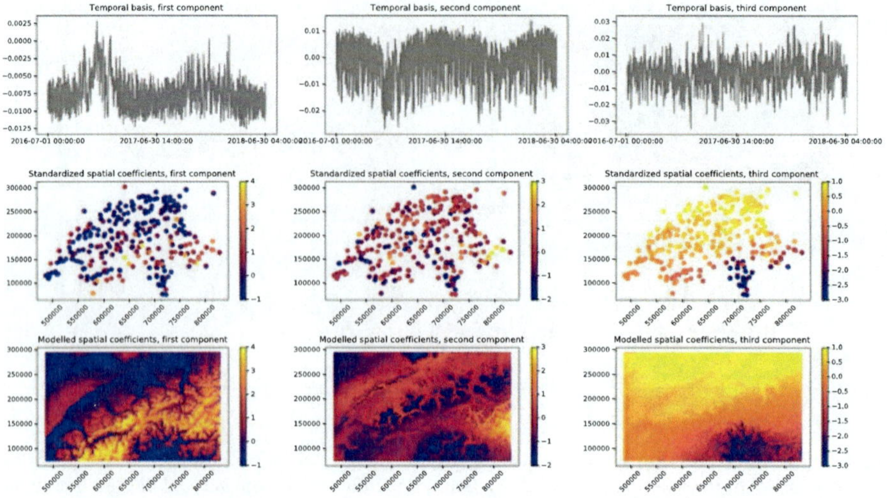

Fig. 5.5 Temperature monitoring network, first 3 components of the EOFs decomposition: (top row) The temporally referenced basis functions; (centre row) The normalised PC spatial coefficients of the corresponding EOFs; (bottom row) The corresponding predicted spatial coefficients provided by the auxiliary outputs of the fully connected neural network (all components)

Starting from measurements sampled with hourly frequency from 1st July 2016 to 30th June 2018 over 369 meteorological stations, the spatio-temporal temperature field will be modelled on a regular grid with a resolution of 2500 [m]. Data were downloaded from the MeteoSwiss website and had a treatment analogous to what was done on wind speed data in Chap. 2. Before modelling, the data set was randomly divided into training, validation and testing subsets, consisting of 220, 75 and 74 stations, respectively.

The first three components resulting from the EOFs decomposition of the training and validation sets, together with the corresponding temporal bases functions and normalised PC spatial coefficients, are displayed in Fig. 5.5. The first two temporal bases clearly show yearly cycles, and a closer exploration of the time series would reveal other structured features such as daily cycles—here not appreciable because of the visualisation of a large number of time-indices. The normalised PC spatial coefficient maps unveil varying patterns at different spatial scales.

Figure 5.6 shows that the first 24 components explain 95% of the data variation. Two different models are implemented. The first one is developed by using all the available components ($K = 294$), while the second one adopts a compressed signal keeping 95 % of data variance ($K = 24$). In addition to latitude and longitude, altitude is added as a spatial covariate in the two models, as temperature strongly depends on it [11]. Therefore, the trained models will have three inputs, i.e. the three spatial covariates and K outputs, with the latter changing in the two experiments conducted. The MAE loss function is used for the optimisation process.

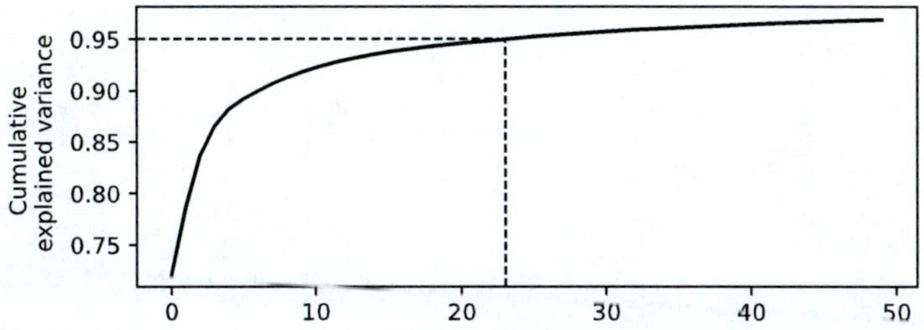

Fig. 5.6 Cumulative percentage of explained variance for the temperature data set. The variance explained by the first 50 components of the EOFs decomposition is displayed. The sum of the relative variance of the first 24 components reaches 95% of the total data variation

Predicted maps of temperature at a randomly chosen fixed time are shown in Fig. 5.7 for both models, together with time series at a random testing station. The predicted temperatures at the testing station are compared with the true measurements on accuracy plots and on time series plot. The model with $K = 294$ replicates extremely well the temperature behaviour. The predicted map captures the different climatic zones, while the predicted time series retrieves the temporal dependencies in the data very well. In particular, highly structured patterns are recovered, such as daily cycles, variability at smaller temporal scales, and abrupt behavioural changes. The accuracy plot further highlights how well the predictions fit the true values. The model with $K = 24$ shows comparable results while the dimensionality of the data has been significantly reduced, indicating that it is possible to obtain similar accuracy with compressed data.

Even if not strictly required to model the spatio-temporal field, the spatial coefficient maps can be obtained from the neural network as auxiliary outputs—shown in Fig. 5.5. Their usage is highly relevant from a diagnostic and interpretation perspective. Indeed, Earth system scientists are accustomed to the use of EOFs as an exploratory data analysis tool to understand the spatio-temporal patterns of atmospheric and environmental phenomena, as exemplified in chapter 3. Hence, the complete reconstruction of the spatial coefficients on a regular grid represents a useful step towards a better explainability of the modelled phenomena. Specifically, these maps could be interpreted from a physical standpoint to analyse the contribution of each temporal variability pattern [5, 12]. In this particular case study, the globally emerging structures correspond to the different known climatic zones. As an example, the first map in the bottom row of Fig. 5.5 clearly shows the contribution of the first temporal basis in the Alps chain, while the third map indicates a substantial negative contribution of the corresponding temporal basis at the south of the chain.

As in the case of the simulated data set, a comparison is made between the multiple-output strategy, where the spatial coefficients are modelled jointly, and the use of separate regression models to predict each coefficient map (Table 5.1). Once more, the latter approach results in a higher error, for both models, with all components and the first 24 ones. Again, using a single network to model the spatial coefficients

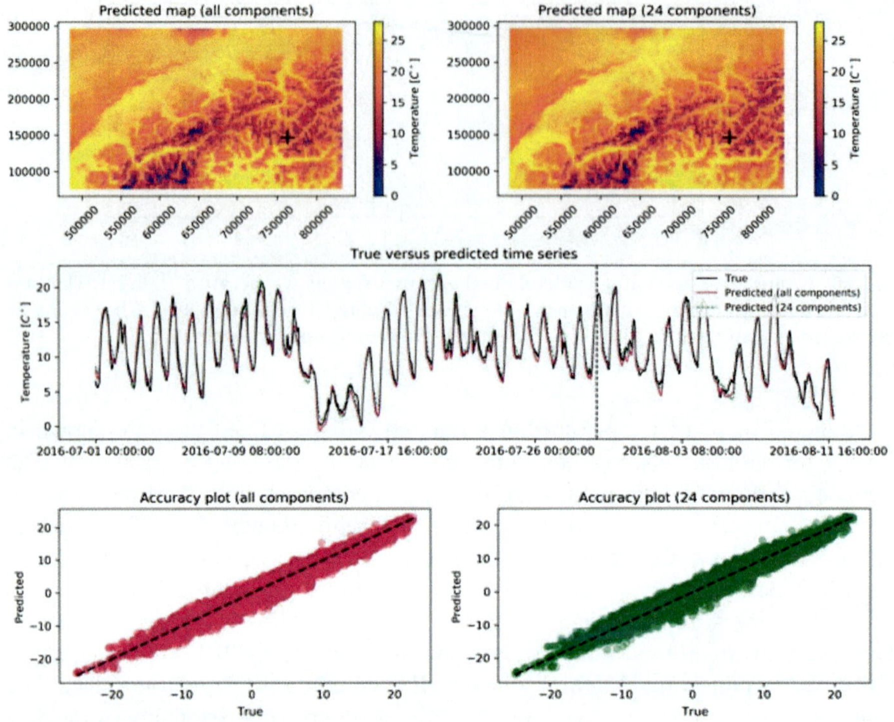

Fig. 5.7 Model prediction for the temperature monitoring network: (top left) The predicted map of temperatures using all EOFs components, at the fixed time indicated by the vertical dashed line in the temporal plot below; (top right) The predicted map using only the first 24 components at the same time; (centre) The true time series (in black) at a testing station marked by a cross in the maps above, the predicted time series with all EOFs components (in magenta) and the predicted time series with the first 24 EOFs components (in green). For visualisation purposes, only the first 42 days of the time series are shown; (bottom left) Accuracy plot at the same testing station for the model with all EOFs components; (bottom right) Accuracy plot at the same testing station for the model with the first 24 EOFs components

and the spatio-temporal fields jointly yields better performances since the algorithm is trained to minimise a loss computed on the final output after the recomposition of the signal.

A variography study is performed on the testing data for the first 42 days. Figure 5.8 shows the spatio-temporal semivariograms for the raw testing data, for the modelled data and residuals resulting from the models, with all the components and with only the first 24 ones. While both models can coherently reconstruct the variability of the raw data, the semivariogram for the model including all the components has a sill comparable to the raw data one. The sill of the semivariogram computed on the modelled data using 24 components is slightly lower, indicating that a certain amount of the variability of the data has not been captured by the model. It may be related to the fact that about 5% of the variability of the training data was not explained

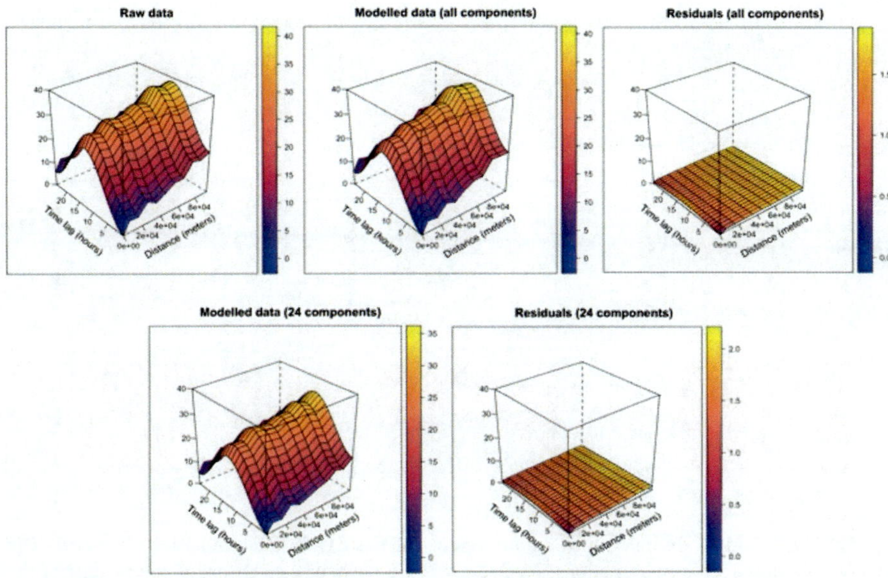

Fig. 5.8 Variography for the temperature data set for the first 42 days. Spatio-temporal semivariograms have been computed on the testing points for the temperature monitoring network data: (top row) Semivariograms for the raw data (left, s^2 = 35.4), the model using all the EOFs components (centre, s^2 = 35.9) and its residuals (right, s^2 = 1.8); (bottom row) Semivariograms for the model using the first 24 EOFs components (left, s^2 = 31.1) and its residuals (right, s^2 = 1.9)

by the first 24 components, as indicated in Fig. 5.6, although other causes could be invoked. The two semivariograms for the residuals show that the models could retrieve most of the spatio-temporal structure of the data. Nonetheless, it can be seen how residuals still show a small temporal correlation corresponding to the daily cycle of temperature. It is likely because the spatial modelling of each component induces an error, which in the reconstructed field becomes proportional to its corresponding temporal basis. It suggests that even if the spatial component models are correctly modelled, some temporal dependencies may subsist.

5.5 Experiment on the MeteoSwiss Data

The proposed framework is finally tested on the MSWind 17 data set presented in Chap. 2 which constitute a more challenging case study due to the nature of wind speed data and the lower number of available training stations—see Table 2.2. The validation set consisting of 40 stations is taken from the training set, which drops to 126 stations. The spatio-temporal wind speed field will be modelled on a 2500 [m] resolution regular grid.

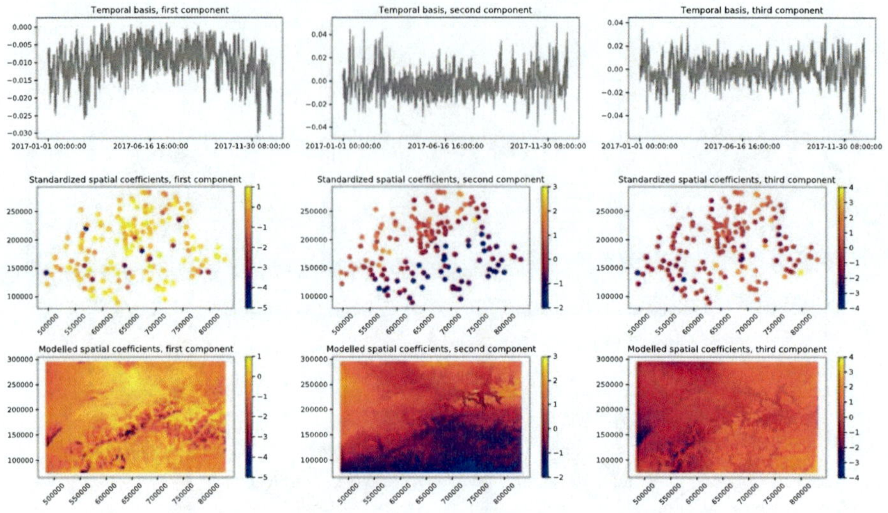

Fig. 5.9 MSWind 17 data set, first 3 components of the EOFs decomposition: (top row) The temporally referenced basis functions; (centre row) The normalised PC spatial coefficients of the corresponding EOFs; (bottom row) The corresponding predicted spatial coefficients provided by the auxiliary outputs of the fully connected neural network (all components)

The first three EOF components of training and validation sets are shown in Fig. 5.9. The first three components explain respectively 47.9, 7.8 and 4.9% of the variability. The patterns of the first two temporal bases and spatial coefficient maps are quite comparable with the results on MSWind 13–16 data set shown in Fig. 3.2.

The same fully connected multiple-output FNN model is trained using all EOF components, and altitude is added to latitude and longitude as spatial covariates. The MSE loss function is selected for the error minimisation to compare with future results of Chap. 7. Root Mean Squared Error (RMSE) computed on the testing set is 1.845. The testing MAE is 1.240.

For a randomly chosen fixed time, a predicted map of wind speed is visualised in Fig. 5.10, as well as a comparison of the true measurements and predicted wind speed through time series and accuracy plots at the BUS (Buchs, 386 [m] of altitude) testing station. The model seems to replicate well the expected behaviour, although it is not perfect. The global trend is reproduced. Some small scale behaviours seem to be somewhat different in some periods, while it is qualitatively similar during other periods. From the 17th of January, the prediction seems to shift over the true values slightly. Note that abrupt behavioural changes and sharp peaks are reasonably well reproduced. The accuracy plot shows a common smoothing effect. It is also quite variable, which could be expected considering that the data under study is the wind speed at an hourly sampling period. The prediction map snapshot as a fixed hour displays a lower wind speed on the Plateau than on the Alps and Jura, which is consistent.

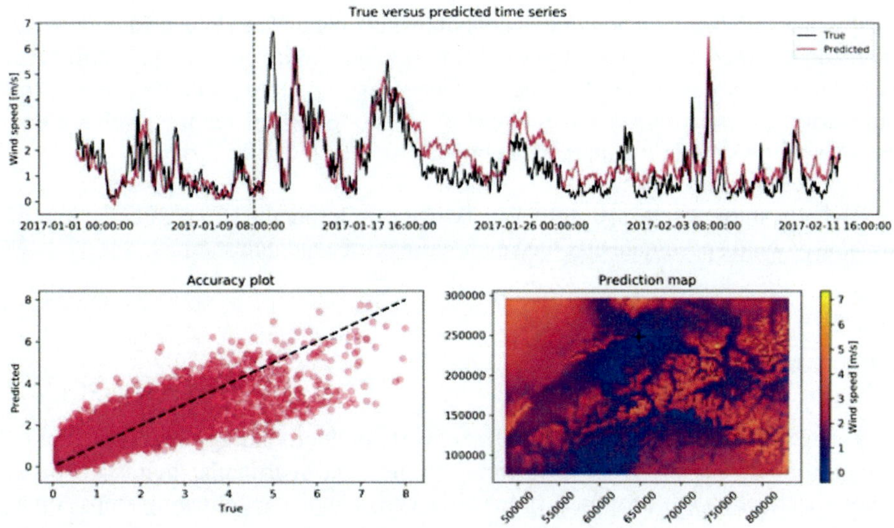

Fig. 5.10 Model prediction for the MSWind 17 data set: (top) The true time series (in black) at the BUS testing station marked by a cross in the maps below and the predicted time series (in magenta). For visualisation purposes, only the first 42 days of the time series are shown; (bottom left) Accuracy plot at the same testing station; (bottom right) The predicted map of wind speed, at the fixed time indicated by the vertical dashed line in the temporal plot above

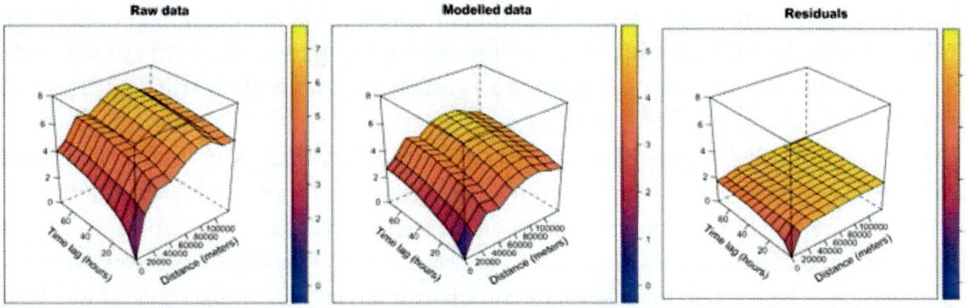

Fig. 5.11 Variography for the MSWind 17 data set for the first 42 days. Spatio-temporal semivariograms have been computed on the training points for the wind speed monitoring network data in 2017: (left) Semivariograms of the raw data ($s^2 = 6.34$); (centre) Semivariograms of the model ($s^2 = 4.42$); (right) Semivariograms of the residuals ($s^2 = 2.21$)

Finally, spatio-temporal semivariograms performed on the training data for the first 42 days are shown in Fig. 5.11, for raw data, model predictions and their residuals. The spatio-temporal structure of raw data is consistently similar to the one shown in Fig. 3.3, January 2017, as the period of studies overlay. The model is able to replicate the spatio-temporal dependencies detected in the raw data by variography. Indeed, the shape of the two semivariograms is very similar. However, the sill is lower for the modelled data. It suggests that the model has lost a substantial amount of variability. It may be partially due to the noisy nature of the raw data or unresolved small scale variability, although it is not completely clear. The semivariogram of the residuals is almost flat. Along the temporal axis, a residual correlation is still here, as in the temperature case. Although a short-scale structure could be present, the semivariogram is close to a pure nugget effect along the spatial axis. Remark that the sum of the sample variance of the modelled data and their residuals is close to the raw data sample variance.

Globally, these results are pretty satisfying considering the complex behaviour of the hourly wind speed data and the small number of available stations.

5.6 Summary

This chapter introduced an ML-based framework for interpolating continuous spatio-temporal fields starting from measurements on a set of irregular points in space. More specifically, the proposed framework consists of the following steps. First, a basis function representation is used to extract fixed temporal bases from spatio-temporal observations. Then, the stochastic spatial coefficients corresponding to each basis function are modelled at any desired spatial location using any ML regression algorithm. Finally, the spatio-temporal signal is recomposed, returning a spatio-temporal interpolation of the field. The proposed approach permits the modelling of non-stationary spatio-temporal processes with multiple cycles and high-frequency behaviour in a non-linear fashion, as the spatial coefficients maps are obtained using non-linear models. In particular, it can interpolate ground measurements of environmental variables keeping into account the spatio-temporal dependencies present in the data.

An example of the framework was specified with an EOF basis and fully connected FNN and tested successfully on synthetic and real-world environmental data. All coefficients are modelled at once, taking advantage of the multiple-output design of the network and learning the spatial structure of each map jointly. Moreover, the loss function can be optimised directly on the reconstructed spatio-temporal field, leading to better results than the alternative way of using a single-output algorithm on each component. However, this alternative way will be investigated in Chap. 7 which will enable variance estimation of the prediction. Each component will be modelled with ELM, which provides a prediction and estimates of its variance. The next chapter is devoted to the development of such UQ methods with ELM.

References

1. Kanevski M, Pozdnoukhov A, Timonin V (2009) Machine learning for spatial environmental data. EPFL Press
2. Goodfellow I, Bengio Y, Courville A (2016) Deep learning. MIT Press

3. Bischke B, Helber P, Folz J, Borth D, Dengel A (2019) Multi-task learning for segmentation of building footprints with deep neural networks. In: 2019 IEEE international conference on image processing (ICIP). IEEE, pp 1480–1484
4. Hooker J, Duveiller G, Cescatti A (2018) A global dataset of air temperature derived from satellite remote sensing and weather stations. Sci Data 5(180246)
5. Wikle C, Zammit-Mangion A, Cressie N (2019) Spatio-temporal Statistics with R, Chapman & Hall/CRC the R series. CRC Press, Taylor & Francis Group
6. Hristopulos DT (2020) Random fields for spatial data modeling. Springer
7. Wikle CK (2019) Comparison of deep neural networks and deep hierarchical models for spatio-temporal data. arXiv:1902.08321 [stat.ML]
8. Abadi M, Agarwal A, Barham P, et al (2015) TensorFlow: large-scale machine learning on heterogeneous systems, Software available from tensorflow.org [Online]. http://tensorflow.org/
9. Amato F, Guignard F, Robert S, Kanevski M (2020) A novel framework for spatio-temporal prediction of environmental data using deep learning. Sci Rep 10(1):22 243. https://doi.org/10.1038/s41598-020-79148-7
10. Schlather M, Malinowski A, Menck PJ, Oesting M, Strokorb K (2015) Analysis, simulation and prediction of multivariate random fields with package random fields. J Stat Softw 63(8):1–25. [Online]. http://www.jstatsoft.org/v63/i08/
11. Whiteman C (2000) Mountain meteorology: fundamentals and applications. Oxford University Press
12. Cressie N, Wikle C (2011) Statistics for Spatio-Temporal Data. Wiley

Chapter 6
Uncertainty Quantification with Extreme Learning Machine

On the one hand, UQ is crucial to assess the prediction quality of an ML model. It is particularly true in the previous chapter setup, where a relatively small number of points are modelled in a high-dimensional space. On the other hand, ELM is a universal approximator for non-linear regression problems that benefit from numerous advantages inherited from linear regression. It is very fast, easy to implement, has very few hyper-parameters and has a one-step optimisation process with a unique analytical solution. This chapter presents original and rigorous UQ developments of ELM. Section 6.1 introduces the context of this development. Section 6.2 briefly reviews ELM theory and sets notations. Starting from general assumptions on the noise covariance matrix, probabilistic formulas are derived for predicted output variance knowing the training input for single and ensemble of (regularised) ELM in Sect. 6.3. Based on these formulas, Sect. 6.4 provides variance estimates when noise is independent with constant variance (homoskedastic case) and non-constant variance (heteroskedastic case). The effectiveness of the proposed estimates is demonstrated through numerical experiments in Sect. 6.5, where the estimation of confidence intervals is also discussed. A scikit-learn compatible Python library enables efficient computation of all estimates discussed herein; see the software availability in Sect. 1.5 of the thesis introduction.

6.1 Related Work and Motivation

Statistical accuracy measures such as variance, standard error, and Confidence Intervals (CI) are crucial to assessing prediction's quality. Model UQ is needed to build CI and has a direct impact on the prediction interval, especially when dealing with small data sets [1]. Uncertainty quantities for FNN solving regression tasks can be

F. Guignard, *On Spatio-Temporal Data Modelling and Uncertainty Quantification Using Machine Learning and Information Theory* , Springer Theses,
https://doi.org/10.1007/978-3-030-95231-0_6

obtained through different methods [2, 3]. Here these quantities are investigated for the ELM model [4]. ELM is a single-layer FNN with random input weights and biases, allowing the output weights' optimisation through the Least-Squares (LS) procedure. One can think about ELM as a projection of inputs in a random feature space where a Multiple Linear Regression (MLR) with a null intercept is performed.

Three main uncertainty sources can be distinguished [5]. The first one comes from the data, particularly from sampling variation and unexplained fluctuations or noise. The second uncertainty source is related to the estimation of the model parameters, which in the case of an FNN correspond to the weights and biases [3]. In ELM, input weights and biases are randomly chosen, which generates uncertainty. Moreover, despite being optimised through a procedure with a unique solution, the estimation of the output weights depends on the random input weights and the data, which induces additional fluctuations. Finally, the third type of uncertainty source is due to the model structure. This source, generally referred to as structural uncertainty, is not considered in this thesis.

Several methods were proposed to obtain confidence or prediction intervals with ELM. A Bayesian formulation was introduced to integrate prior knowledge and produce CI directly [6, 7]. In the frequentist paradigm, bootstrap methods were investigated in the context of time series [8]. Akusok et al. proposed a method to estimate prediction intervals using a covariance matrix estimate coming from MLR [9].

Most of these methods make a Gaussian assumption on the output distribution or do not consider the bias in confidence/prediction interval estimation, which may cause misleading conclusions. Moreover, resampling methods lead to substantial computational burdens when the number of data is high. Finally, it is often argued that the randomness of the input weights and biases is supposed to be negligible, providing the training set is large enough. However, it is not always clear how many data is needed in practice. Indeed, while stochastic input layer initialisation can have a weak impact in low dimension, it is still unclear what could happen when the number of features or/and neurons are large. Because of the curse of dimensionality, the random drawing of the input weights and biases could have a higher impact than suspected. Therefore, the development of variance estimation methods considering the ELM stochastic nature is highly relevant to investigate such impact further. Additionally, ELM is also used efficiently with small training data set [10]—in which case precise variance estimate is crucial—where the randomness of the input weights and biases should not be ignored.

6.2 Background and Notations

Let us state the regression problem while the notation is fixed. Assume that an output variable y depends on d input variables $x_1, \ldots, x_d$ through the relationship

$$y = f(\mathbf{x}) + \varepsilon(\mathbf{x}),$$

where $\mathbf{x} = (x_1, \ldots, x_d)^T \in \mathbb{R}^d$ is the vector composed by the input variables, f is a function of $\mathbf{x}$ whose value represents the deterministic part of y, and $\varepsilon(\mathbf{x})$ is a random noise depending on the input representing the stochastic part of y. It is assumed that, whatever the value of $\mathbf{x}$, the noise is centred and has a finite variance.

Let the training set $\mathcal{D} = \{(\mathbf{x}_i, y_i) : \mathbf{x}_i \in \mathbb{R}^d, y_i \in \mathbb{R}\}_{i=1}^n$ be a sample from the joint distribution of $(\mathbf{x}, y)$. Given a new input point $\mathbf{x}_0 \in \mathbb{R}^d$, one wants to predict the corresponding output y_0. The value $f(\mathbf{x}_0)$ seems a good guess. However, the function f is unknown. The true function f needs to be approximated based on the sample $\mathcal{D}$ to provide an estimate of the prediction $f(\mathbf{x}_0)$.

For convenience, the matrix composed by all training input points will be noted $\mathbf{X} = [\mathbf{x}_1 \mid \cdots \mid \mathbf{x}_n] \in \mathbb{R}^{d \times n}$. Moreover, at the training points, the $n-$dimensional vectors $\mathbf{y} = (y_1, \ldots, y_n)^T$, $\mathbf{f} = (f(\mathbf{x}_1), \ldots, f(\mathbf{x}_n))^T$, and $\boldsymbol{\varepsilon} = (\varepsilon(\mathbf{x}_1), \ldots, \varepsilon(\mathbf{x}_n))^T$ are defined. The covariance matrix of $\boldsymbol{\varepsilon}$ knowing $\mathbf{X}$ is noted $\boldsymbol{\Sigma} \in \mathbb{R}^{n \times n}$.

6.2.1 Extreme Learning Machine

Extreme Learning Machine (ELM) is a single-layer FNN with random initialisation of the input weights $\mathbf{w}_j \in \mathbb{R}^d$ and biases $b_j \in \mathbb{R}$, for $j = 1, \ldots N$, where N denotes the number of neurons of the hidden layer. All input weights and biases are i.i.d. and are generally sampled from a Gaussian or uniform distribution. They map the input space into random feature space in a non-linear fashion by-way-of the *non-linear feature mapping* [11]

$$\mathbf{h}(\mathbf{x}) = \left(h_1(\mathbf{x}), \ldots, h_N(\mathbf{x})\right)^T \in \mathbb{R}^N,$$

where $h_j(\mathbf{x}) = g(\mathbf{x}^T \mathbf{w}_j + b_j)$ is the output of the jth hidden node, for $j = 1, \ldots, N$, and g is any infinitely differentiable activation function [12]. An $N-$hidden neurons ELM can generate output functions of the form

$$f_N(\mathbf{x}) = \mathbf{h}(\mathbf{x})^T \boldsymbol{\beta},$$

where $\boldsymbol{\beta} \in \mathbb{R}^N$ is the vector of output weights that relate the hidden layer with the output node.

The output weights are trained using the sample $\mathcal{D}$ and optimised regarding the L_2 criterion performing the following procedure. The $n \times N$ hidden layer matrix, denoted $\mathbf{H}$ and defined element-wise by $\mathbf{H}_{ij} = h_j(\mathbf{x}_i)$, $i = 1, \ldots, n$, and $j = 1, \ldots, N$, is computed. Then the cost function J_1, defined by

$$J_1(\boldsymbol{\beta}) = ||\mathbf{y} - \mathbf{H}\boldsymbol{\beta}||_2^2,$$

where $|| \cdot ||_2$ denotes the Euclidean norm, is minimised. This is exactly the LS procedure for a design matrix $\mathbf{H}$ [13, 14]. If the matrix $\mathbf{H}^T\mathbf{H}$ is of full rank and then invertible, the output weights are estimated as the classical linear regression, with the analytical solution of the minimisation of $J_1(\boldsymbol{\beta})$,

$$\widehat{\boldsymbol{\beta}} = (\mathbf{H}^T\mathbf{H})^{-1}\mathbf{H}^T\mathbf{y},$$

where the matrix $(\mathbf{H}^T\mathbf{H})^{-1}\mathbf{H}^T$ is the Moore-Penrose generalised inverse of the matrix $\mathbf{H}$ and will be denoted $\mathbf{H}^\dagger$ in the following. Thus, ELM can be thought of as an MLR with a null intercept, performed on regressors obtained by a random non-linear transformation of the input variables.

At a new point, the prediction is given by $\hat{f}(\mathbf{x}_0) = \mathbf{h}(\mathbf{x}_0)^T\widehat{\boldsymbol{\beta}}$. In the remainder of the chapter, all dependencies in $\mathbf{x}_0$ will be dropped for convenience and the prediction will be noted $\hat{f}(\mathbf{x}_0) = \mathbf{h}^T\widehat{\boldsymbol{\beta}} = \mathbf{h}^T\mathbf{H}^\dagger\mathbf{y}$. The vector of model predictions at training points will be noted $\hat{\mathbf{f}}$, defined element-wise by $\hat{\mathbf{f}}_i = \hat{f}(\mathbf{x}_i)$. The (random) matrix of all input weights and biases will be denoted

$$\mathbf{W} = \left[\begin{array}{c|c|c} \mathbf{w}_1 & \ldots & \mathbf{w}_N \\ \hline b_1, & \ldots & b_N, \end{array}\right].$$

6.2.2 *Regularised ELM*

To avoid overfitting and reduce outlier effects, a regularised version of ELM was proposed [15]. Highly variable output weights due to multicollinearity among neurons can be stabilised with regularisation, too. As mentioned by [16], this model is basically a particular case of *Tikhonov regularisation*—also known as *ridge regression* [17]—performed on the random feature space. The output weights are optimised regarding the cost function J_2, where

$$J_2(\boldsymbol{\beta}) = ||\mathbf{y} - \mathbf{H}\boldsymbol{\beta}||_2^2 + \alpha||\boldsymbol{\beta}||_2^2,$$

for some real number $\alpha > 0$, sometimes called the *Tikhonov factor*, which controls the penalisation of taking big output weights. The analytical solution of this optimisation problem for a fixed α is given by

$$\widehat{\boldsymbol{\beta}} = (\mathbf{H}^T\mathbf{H} + \alpha\mathbf{I})^{-1}\mathbf{H}^T\mathbf{y},$$

thanks to the fact that the matrix $(\mathbf{H}^T\mathbf{H} + \alpha\mathbf{I})$ is always invertible; see [18]. To lighten the notation, the matrix $\mathbf{H}^\alpha = (\mathbf{H}^T\mathbf{H} + \alpha\mathbf{I})^{-1}\mathbf{H}^T$ is defined. Remark that as α goes to zero, $\mathbf{H}^\alpha$ goes to $\mathbf{H}^\dagger$ and the classical ELM is recovered [15]. In the remainder of the book, most of the results are presented with $\mathbf{H}^\alpha$, but are still valid for the non-regularised case unless the contrary is specified.

6.2.3 ELM Ensemble

Another way to avoid overfitting is to combine several ELM models. It also reduces the randomness induced by the input weight initialisation, which could be beneficial—especially for small data sets. Several ensemble techniques have been developed for ELM [16, 19]. In this book, each model of a given ensemble will have the same activation function and number of neurons, and all models will be averaged after training. It corresponds to retrain M times the model with different input weight initialisation and average the results, where M is the number of ELM networks in the ensemble. The hidden layer matrix and the matrix of input weights and biases of the mth retraining will be noted respectively $\mathbf{H}_m$ and $\mathbf{W}_m$, for $m = 1, \ldots, M$. If the mth prediction is noted $\hat{f}_m(\mathbf{x}_0)$, the final prediction is

$$\hat{f}(\mathbf{x}_0) = \frac{1}{M}\sum_{m=1}^{M} \hat{f}_m(\mathbf{x}_0) = \frac{1}{M}\sum_{m=1}^{M} \mathbf{h}_m^T \mathbf{H}_m^\alpha \mathbf{y},$$

where $\mathbf{h}_m$ and $\mathbf{H}_m^\alpha$ are the analogous quantities defined previously for the mth model. Note that the input weights are drawn from the same distribution, hence have the same joint distribution across all the models, which allows us to drop the m index in most calculations of the remainder of this chapter.

6.3 Analytical Developments

This section introduces new analytical results on ELM. It begins with the derivation of the bias and variance for a single (regularised) ELM. Subsequently, the results are generalised to ELM ensembles. Finally, the correlation between two ELMs is investigated.

6.3.1 Bias and Variance for a Single ELM

For all derivations, it is supposed that hyper-parameters N and α—when appropriate—are considered as fixed and non-stochastic. Also, all formulas are derived knowing $\mathbf{X}$. However, this conditioning is dropped to avoid cumbersome notations. Note that $\hat{f}(\mathbf{x}_0)$ is a random variable depending on the noise at training points ε, but also on the input weights and biases $\mathbf{W}$ used in the construction of $\mathbf{h}$ and $\mathbf{H}^\alpha$. As the noise is centred,

$$\mathbb{E}\left[\hat{f}(\mathbf{x}_0) \,\middle|\, \mathbf{W}\right] = \mathbf{h}^T \mathbf{H}^\alpha \mathbb{E}\left[\mathbf{f} + \varepsilon\right] = \mathbf{h}^T \mathbf{H}^\alpha \mathbf{f}. \tag{6.1}$$

Using the law of total expectation, one can compute the bias of the model at $\mathbf{x}_0$,

$$\text{Bias}\left[\hat{f}(\mathbf{x}_0)\right] = \mathbb{E}\left[\mathbb{E}\left[\hat{f}(\mathbf{x}_0)\,\middle|\,\mathbf{W}\right]\right] - f(\mathbf{x}_0) = \mathbb{E}\left[\mathbf{h}^T\mathbf{H}^\alpha\mathbf{f}\right] - f(\mathbf{x}_0).$$

Let us now compute the variance of the model at a new point. First, one has

$$\text{Var}\left[\hat{f}(\mathbf{x}_0)\,\middle|\,\mathbf{W}\right] = \mathbf{h}^T\mathbf{H}^\alpha \text{Var}\left[\mathbf{f}+\varepsilon\right]\mathbf{H}^{\alpha T}\mathbf{h} = \mathbf{h}^T\mathbf{H}^\alpha\mathbf{\Sigma}\mathbf{H}^{\alpha T}\mathbf{h}, \tag{6.2}$$

which is the typical variance expression for MLR. With Eqs. (6.1) and (6.2), the variance of the model at $\mathbf{x}_0$ can be computed by using the law of total variance,

$$\begin{aligned}\text{Var}\left[\hat{f}(\mathbf{x}_0)\right] &= \mathbb{E}\left[\text{Var}\left[\hat{f}(\mathbf{x}_0)\,\middle|\,\mathbf{W}\right]\right] + \text{Var}\left[\mathbb{E}\left[\hat{f}(\mathbf{x}_0)\,\middle|\,\mathbf{W}\right]\right] \\ &= \mathbb{E}\left[\mathbf{h}^T\mathbf{H}^\alpha\mathbf{\Sigma}\mathbf{H}^{\alpha T}\mathbf{h}\right] + \text{Var}\left[\mathbf{h}^T\mathbf{H}^\alpha\mathbf{f}\right].\end{aligned} \tag{6.3}$$

The first term of RHS of the Eq. (6.3) is the variance of the LS step averaged on all possible random feature spaces generated by input weights and biases, while the second term is the variation of the LS step bias across all random feature spaces. Note that the second term appears if and only if the random input weights and biases are considered. The LS step bias appears since that the data are not assumed to be distributed around a linear subspace. In the non-regularised case with independent homoskedastic noise, if $\mathbf{W}$ and $\mathbf{X}$ are deterministic, the classical MLR formula for the variance at a prediction point is recovered; see [14].

6.3.2 *Bias and Variance for ELM Ensemble*

As mentioned before, the training could be done several times and averaged. The bias and variance of the averaged predictor are obtained in this subsection by a direct calculation using the law of total variance and elementary probability calculus. Recall that $\mathbf{W}_1, \ldots \mathbf{W}_M$ are i.i.d.. Reusing Eq. (6.1), one has

$$\begin{aligned}\mathbb{E}\left[\hat{f}(\mathbf{x}_0)\,\middle|\,\mathbf{W}_1,\ldots,\mathbf{W}_M\right] &= \frac{1}{M}\sum_{m=1}^{M}\mathbb{E}\left[\hat{f}_m(\mathbf{x}_0)\,\middle|\,\mathbf{W}_m\right] \\ &= \frac{1}{M}\sum_{m=1}^{M}\mathbf{h}_m^T\mathbf{H}_m^\alpha\mathbf{f}.\end{aligned} \tag{6.4}$$

The law of total expectation yields

$$\mathbb{E}\left[\hat{f}(\mathbf{x}_0) - f(\mathbf{x}_0)\right] = \mathbb{E}\left[\mathbb{E}\left[\hat{f}(\mathbf{x}_0) \,\middle|\, \mathbf{W}_1, \ldots, \mathbf{W}_M\right]\right] - f(\mathbf{x}_0)$$
$$= \frac{1}{M}\sum_{m=1}^{M} \underbrace{\mathbb{E}\left[\mathbf{h}_m^T \mathbf{H}_m^{\alpha} \mathbf{f}\right]}_{=\text{cst}} - f(\mathbf{x}_0)$$
$$= \mathbb{E}\left[\mathbf{h}^T \mathbf{H}^{\alpha} \mathbf{f}\right] - f(\mathbf{x}_0),$$

and the bias still unchanged. Computing the first term of the law of total variance, one gets

$$\mathrm{Var}\left[\hat{f}(\mathbf{x}_0) \,\middle|\, \mathbf{W}_1, \ldots, \mathbf{W}_M\right] = \mathrm{Var}\left[\frac{1}{M}\sum_{m=1}^{M} \mathbf{h}_m^T \mathbf{H}_m^{\alpha} \mathbf{y} \,\middle|\, \mathbf{W}_1, \ldots, \mathbf{W}_M\right]$$
$$= \frac{1}{M^2}\sum_{m,l=1}^{M} \mathrm{Cov}\left[\mathbf{h}_m^T \mathbf{H}_m^{\alpha} \mathbf{y},\ \mathbf{h}_l^T \mathbf{H}_l^{\alpha} \mathbf{y} \,\middle|\, \mathbf{W}_m, \mathbf{W}_l\right]$$
$$= \frac{1}{M^2}\sum_{m,l=1}^{M} \mathbf{h}_m^T \mathbf{H}_m^{\alpha} \mathrm{Var}\left[\mathbf{y}\right] \mathbf{H}_l^{\alpha T} \mathbf{h}_l$$
$$= \frac{1}{M^2}\sum_{m,l=1}^{M} \mathbf{h}_m^T \mathbf{H}_m^{\alpha} \boldsymbol{\Sigma} \mathbf{H}_l^{\alpha T} \mathbf{h}_l,$$

which implies

$$\mathbb{E}\left[\mathrm{Var}\left[\hat{f}(\mathbf{x}_0) \,\middle|\, \mathbf{W}_1, \ldots, \mathbf{W}_M\right]\right] = \frac{1}{M^2}\sum_{m=1}^{M} \underbrace{\mathbb{E}\left[\mathbf{h}_m^T \mathbf{H}_m^{\alpha} \boldsymbol{\Sigma} \mathbf{H}_m^{\alpha T} \mathbf{h}_m\right]}_{=\text{cst}}$$
$$+ \frac{1}{M^2}\sum_{m \neq l} \mathbb{E}\left[\mathbf{h}_m^T \mathbf{H}_m^{\alpha}\right] \boldsymbol{\Sigma} \mathbb{E}\left[\mathbf{H}_l^{\alpha T} \mathbf{h}_l\right] \tag{6.5}$$
$$= \frac{1}{M}\mathbb{E}\left[\mathbf{h}^T \mathbf{H}^{\alpha} \boldsymbol{\Sigma} \mathbf{H}^{\alpha T} \mathbf{h}\right]$$
$$+ \frac{M-1}{M}\mathbb{E}\left[\mathbf{h}^T \mathbf{H}^{\alpha}\right] \boldsymbol{\Sigma} \mathbb{E}\left[\mathbf{H}^{\alpha T} \mathbf{h}\right],$$

where one used the fact that the input weights and biases are drawn independently. Using (6.4), the second term of the law of total variance is

$$\begin{aligned}\mathrm{Var}\left[\mathbb{E}\left[\hat{f}(\mathbf{x}_0)\,\middle|\,\mathbf{W}_1,\ldots,\mathbf{W}_M\right]\right] &= \frac{1}{M^2}\sum_{m,l=1}^{M}\mathrm{Cov}\left[\mathbf{h}_m^T\mathbf{H}_m^\alpha\mathbf{f},\ \mathbf{h}_l^T\mathbf{H}_l^\alpha\mathbf{f}\right]\\ &= \frac{1}{M^2}\sum_{m=1}^{M}\underbrace{\mathrm{Var}\left[\mathbf{h}_m^T\mathbf{H}_m^\alpha\mathbf{f}\right]}_{=\,\mathrm{cst}}\\ &= \frac{1}{M}\mathrm{Var}\left[\mathbf{h}^T\mathbf{H}^\alpha\mathbf{f}\right],\end{aligned} \tag{6.6}$$

as $\mathrm{Cov}\left[\mathbf{h}_m^T\mathbf{H}_m^\alpha\mathbf{f},\ \mathbf{h}_l^T\mathbf{H}_l^\alpha\mathbf{f}\right]$ vanishes when $m \neq l$, thanks again to the i.i.d. assumption on the weights. By summing Eqs. (6.5) and (6.6), the result is obtained and one gets

$$\begin{aligned}\mathrm{Var}\left[\hat{f}(\mathbf{x}_0)\right] = \frac{1}{M}\mathbb{E}\left[\mathbf{h}^T\mathbf{H}^\alpha\boldsymbol{\Sigma}\mathbf{H}^{\alpha T}\mathbf{h}\right] + \frac{M-1}{M}\mathbb{E}\left[\mathbf{h}^T\mathbf{H}^\alpha\right]\boldsymbol{\Sigma}\,\mathbb{E}\left[\mathbf{H}^{\alpha T}\mathbf{h}\right]\\ + \frac{1}{M}\mathrm{Var}\left[\mathbf{h}^T\mathbf{H}^\alpha\mathbf{f}\right].\end{aligned} \tag{6.7}$$

The RHS first and third terms are the single ELM variance divided by the number of models. The bias variation of the LS step is reduced by a $1/M$ factor. Although the average variance of the LS step represented by the RHS first term seems to decrease by a $1/M$ factor, models are pairwise dependent, which yields the RHS second term. Notice that if $M = 1$, Eq. (6.3) is recovered. If M grows, the RHS second term tends to dominate the model variance. Remark also that using the law of total covariance, it is easily checked by an analogous computation that the covariance between two members of an ELM ensemble corresponds to $\mathbb{E}\left[\mathbf{h}^T\mathbf{H}^\alpha\right]\boldsymbol{\Sigma}\,\mathbb{E}\left[\mathbf{H}^{\alpha T}\mathbf{h}\right]$.

6.3.3 *Use of Random Variable Quadratic Forms*

Formulation of variance in Eqs. (6.3) and (6.7) is convenient for the interpretation of ELM as an MLR on random features. However, quadratic forms in random variables appear in these formulas, which allows pursuing calculations. With the corollary 3.2b.1 of [20], the expectation of random variable quadratic form can be computed as the quadratic form in its expected values plus the trace of its covariance matrix times the matrix of the quadratic form. This corollary will be used extensively in the remainder of this chapter, each time an expectation of a quadratic form in random variables appears.

Setting the random vector $\boldsymbol{\xi} = \mathbf{H}^{\alpha T}\mathbf{h}$ and assuming the existence of its expectation $\boldsymbol{\mu}$ and covariance matrix $\mathbf{C}$, Eq. (6.3) becomes

$$\begin{aligned}\mathrm{Var}\left[\hat{f}(\mathbf{x}_0)\right] &= \mathbb{E}\left[\boldsymbol{\xi}^T\boldsymbol{\Sigma}\boldsymbol{\xi}\right] + \mathrm{Var}\left[\mathbf{f}^T\boldsymbol{\xi}\right]\\ &= \mathrm{Tr}\left[\boldsymbol{\Sigma}\mathbf{C}\right] + \boldsymbol{\mu}^T\boldsymbol{\Sigma}\boldsymbol{\mu} + \mathbf{f}^T\mathbf{C}\mathbf{f},\end{aligned}$$

where Tr [·] denotes the trace of a square matrix. Although the notation does not specify it, the quantities $\boldsymbol{\xi}$, $\boldsymbol{\mu}$ and $\mathbf{C}$ depend on the Tikhonov factor α in the regularised case. Similarly, $\boldsymbol{\xi}_m = \mathbf{H}_m^{\alpha T}\mathbf{h}_m$ is set for ELM ensembles and the variance becomes

$$\mathrm{Var}\left[\hat{f}(\mathbf{x}_0)\right] = \frac{1}{M}\mathrm{Tr}\left[\boldsymbol{\Sigma}\mathbf{C}\right] + \boldsymbol{\mu}^T\boldsymbol{\Sigma}\boldsymbol{\mu} + \frac{1}{M}\mathbf{f}^T\mathbf{C}\mathbf{f}. \tag{6.8}$$

6.3.4 Correlation Between Two ELMs

As the covariance between two single ELMs $\hat{f}_1(\mathbf{x}_0)$ and $\hat{f}_2(\mathbf{x}_0)$ is $\boldsymbol{\mu}^T\boldsymbol{\Sigma}\boldsymbol{\mu}$, their linear correlation at $\mathbf{x}_0$ is given by

$$\mathrm{Corr}\left[\hat{f}_1(\mathbf{x}_0),\ \hat{f}_2(\mathbf{x}_0)\right] = \frac{\boldsymbol{\mu}^T\boldsymbol{\Sigma}\boldsymbol{\mu}}{\mathrm{Tr}\left[\boldsymbol{\Sigma}\mathbf{C}\right] + \boldsymbol{\mu}^T\boldsymbol{\Sigma}\boldsymbol{\mu} + \mathbf{f}^T\mathbf{C}\mathbf{f}}.$$

Remark that considering the input weights and biases as fixed is equivalent to ignore $\mathrm{Tr}\left[\boldsymbol{\Sigma}\mathbf{C}\right] + \mathbf{f}^T\mathbf{C}\mathbf{f}$ and to having a correlation of 1 between the two models.

An interesting insight is provided by the case of independent and homoskedastic noise, i.e. $\boldsymbol{\Sigma} = \sigma_\varepsilon^2\mathbf{I}$ with $\sigma_\varepsilon^2 \in \mathbb{R}$, which yields

$$\mathrm{Corr}\left[\hat{f}_1(\mathbf{x}_0),\ \hat{f}_2(\mathbf{x}_0)\right] = \frac{\boldsymbol{\mu}^T\boldsymbol{\mu}}{\mathrm{Tr}\left[\mathbf{C}\right] + \boldsymbol{\mu}^T\boldsymbol{\mu} + \frac{1}{\sigma_\varepsilon^2}\mathbf{f}^T\mathbf{C}\mathbf{f}}.$$

Notice that in this particular case, when σ_ε^2 is small the linear correlation between two ELMs vanishes. Contrariwise, when σ_ε^2 is large the linear correlation between two ELMs tends to $b = \boldsymbol{\mu}^T\boldsymbol{\mu}/(\boldsymbol{\mu}^T\boldsymbol{\mu} + \mathrm{Tr}\left[\mathbf{C}\right])$. Therefore, the amount of noise has a direct impact on the linear correlation which takes its value between 0 and $b \leq 1$. The trace of $\mathbf{C}$ can be interpreted as an uncertainty measure quantifying the variability of $\boldsymbol{\xi}$, called sometimes the *total variation* or the *total dispersion* of $\boldsymbol{\xi}$ [21]. In our case, it controls the linear correlation bound b, in the sense that more variable is $\boldsymbol{\xi}$, farther from 1 is the maximal value that the linear correlation can take, regardless of the noise in the data.

6.4 Variance Estimation of ELM Ensemble

This section introduces novel estimates of the ELM variance. Although several ELMs are necessary to allow the estimation of quantities related to $\boldsymbol{\xi}$—which motivates the use of ELM ensembles—reliable results are also obtained with very small ensembles. First, the variation of the LS step bias overall random feature space is estimated. Then, assuming noise independence, the variance of the LS step averaged on all possible random feature spaces is estimated under homoskedasticity and heteroskedasticity for non-regularised and regularised ELM ensembles.

6.4.1 Estimation of the Least-Squares Bias Variation

The quantity $\mathbf{f}^T\mathbf{C}\mathbf{f} = \mathrm{Var}\left[\mathbf{h}^T\mathbf{H}^\alpha\mathbf{f}\right]$ does not depend on noise. It is the variance induced by the randomness of $\mathbf{W}$, knowing the true function f at training points. Tentatively assume that the output weights are not regularised. As $\mathbf{f}$ is unknown, one approximate it by the model prediction at the training points $\hat{\mathbf{f}} = \mathbf{H}\mathbf{H}^\dagger\mathbf{y}$. For each model of the ensemble,

$$\mathrm{Var}\left[\mathbf{h}^T\mathbf{H}^\dagger\mathbf{f}\right] \approx \mathrm{Var}\left[\mathbf{h}^T\mathbf{H}^\dagger\hat{\mathbf{f}} \,\middle|\, \varepsilon\right] = \mathrm{Var}\left[\hat{f}(\mathbf{x}_0) \,\middle|\, \varepsilon\right].$$

This motivates the following estimate for $\mathbf{f}^T\mathbf{C}\mathbf{f}$,

$$\hat{\sigma}_{\hat{f}}^2 = \frac{1}{M-1}\sum_{m=1}^{M}\left(\hat{f}_m(\mathbf{x}_0) - \frac{1}{M}\sum_{l=1}^{M}\hat{f}_l(\mathbf{x}_0)\right)^2. \tag{6.9}$$

The same estimate will be used for the regularised case. In both cases, this estimate is more than reasonable, as it is shown below in the remainder of this subsection.

The expectation of $\hat{\sigma}_{\hat{f}}^2$ can be easily computed. Knowing the noise at the training points,

$$\begin{aligned}
\mathbb{E}\left[\hat{\sigma}_{\hat{f}}^2 \,\middle|\, \varepsilon\right] &= \mathrm{Var}\left[\hat{f}(\mathbf{x}_0) \,\middle|\, \varepsilon\right] = \mathrm{Var}\left[\boldsymbol{\xi}^T\mathbf{y} \,\middle|\, \varepsilon\right] \\
&= \mathrm{Var}\left[\boldsymbol{\xi}^T\mathbf{f} \,\middle|\, \varepsilon\right] + \mathrm{Var}\left[\boldsymbol{\xi}^T\varepsilon \,\middle|\, \varepsilon\right] + 2\,\mathrm{Cov}\left[\boldsymbol{\xi}^T\mathbf{f},\ \boldsymbol{\xi}^T\varepsilon \,\middle|\, \varepsilon\right] \\
&= \mathrm{Var}\left[\boldsymbol{\xi}^T\mathbf{f}\right] + \varepsilon^T\mathbf{C}\varepsilon + 2\,\mathbf{f}^T\mathbf{C}\varepsilon,
\end{aligned}$$

using the unbiasedness of the estimate in the first equality. By taking the expectation over ε on both side,

$$\mathbb{E}\left[\hat{\sigma}_{\hat{f}}^2\right] = \mathrm{Var}\left[\mathbf{h}^T\mathbf{H}^\alpha\mathbf{f}\right] + \mathrm{Tr}\left[\boldsymbol{\Sigma}\mathbf{C}\right].$$

It shows that the bias of the estimate defined in Eq. (6.9) is given by $\mathrm{Tr}\left[\boldsymbol{\Sigma}\mathbf{C}\right]$, which is the first term of the RHS of Eq. (6.8) up to a factor $1/M$. Therefore, regardless of a particular form of $\boldsymbol{\Sigma}$ or whether ELM is regularised or not, it is unnecessary to estimate the latter and $(1/M)\,\hat{\sigma}_{\hat{f}}^2$ is an unbiased estimate of the sum of the first and last terms of Eq. (6.8).

6.4.2 Estimation Under Independence and Homoskedastic Assumptions

Only the second term of RHS of Eq. (6.8) remains to be estimated. If the noise is assumed to be independent and have a constant variance, the covariance matrix of ε writes $\boldsymbol{\Sigma} = \sigma_\varepsilon^2 \mathbf{I}$ and the second term of RHS of Eq. (6.8) becomes $\boldsymbol{\mu}^T \boldsymbol{\Sigma} \boldsymbol{\mu} = \sigma_\varepsilon^2 \boldsymbol{\mu}^T \boldsymbol{\mu}$. The quantity $\boldsymbol{\mu}^T \boldsymbol{\mu}$—which, knowing $\mathbf{X}$, stochastically depends only on $\mathbf{W}$—will be estimated separately from σ_ε^2.

6.4.2.1 Estimation of $\boldsymbol{\mu}^T \boldsymbol{\mu}$

As a first step, $\boldsymbol{\mu}^T \boldsymbol{\mu}$ is naively estimated with

$$\widehat{\boldsymbol{\mu}}^T \widehat{\boldsymbol{\mu}} = \frac{1}{M^2} \sum_{m,l=1}^{M} \boldsymbol{\xi}_m^T \boldsymbol{\xi}_l, \quad \text{where} \quad \widehat{\boldsymbol{\mu}} = \frac{1}{M} \sum_{m=1}^{M} \boldsymbol{\xi}_m. \tag{6.10}$$

However, remark that

$$\begin{aligned} M^2 \mathbb{E}\left[\widehat{\boldsymbol{\mu}}^T \widehat{\boldsymbol{\mu}}\right] &= \sum_{m=1}^{M} \mathbb{E}\left[\boldsymbol{\xi}_m^T \boldsymbol{\xi}_m\right] + \sum_{m \neq l} \mathbb{E}\left[\boldsymbol{\xi}_m\right]^T \mathbb{E}\left[\boldsymbol{\xi}_l\right] \\ &= M\left(\mathrm{Tr}\left[\mathbf{C}\right] + \boldsymbol{\mu}^T \boldsymbol{\mu}\right) + M(M-1)\boldsymbol{\mu}^T \boldsymbol{\mu} \\ &= M\, \mathrm{Tr}\left[\mathbf{C}\right] + M^2\, \boldsymbol{\mu}^T \boldsymbol{\mu}, \end{aligned}$$

and dividing by M^2 shows that the estimate given in (6.10) has a bias equal to $(1/M)\mathrm{Tr}\left[\mathbf{C}\right]$, which comes from the expected values of the M quadratic terms in $\boldsymbol{\xi}_m$.

To remove this bias, one can estimate it by

$$\frac{1}{M} \mathrm{Tr}[\widehat{\mathbf{Q}}], \quad \text{where} \quad \widehat{\mathbf{Q}} = \frac{1}{M-1} \sum_{m=1}^{M} \left(\boldsymbol{\xi}_m - \widehat{\boldsymbol{\mu}}\right)\left(\boldsymbol{\xi}_m - \widehat{\boldsymbol{\mu}}\right)^T. \tag{6.11}$$

This is an unbiased estimate of $(1/M)\mathrm{Tr}\left[\mathbf{C}\right]$, which immediately follows from the fact that $\widehat{\mathbf{Q}}$ is an unbiased estimate of $\mathbf{C}$. Therefore, subtract (6.11) from (6.10) yields an unbiased estimate of $\boldsymbol{\mu}^T \boldsymbol{\mu}$. Note that

$$\begin{aligned} (M-1)\,\mathrm{Tr}[\widehat{\mathbf{Q}}] &= \sum_{m=1}^{M} \left(\boldsymbol{\xi}_m - \widehat{\boldsymbol{\mu}}\right)^T \left(\boldsymbol{\xi}_m - \widehat{\boldsymbol{\mu}}\right) \\ &= \sum_{m=1}^{M} \boldsymbol{\xi}_m^T \boldsymbol{\xi}_m - M\, \widehat{\boldsymbol{\mu}}^T \widehat{\boldsymbol{\mu}}. \end{aligned}$$

Hence, the unbiased estimate of $\boldsymbol{\mu}^T\boldsymbol{\mu}$ that was just developed results in

$$\begin{aligned}\widehat{\boldsymbol{\mu}}^T\widehat{\boldsymbol{\mu}} - \frac{1}{M}\mathrm{Tr}[\widehat{\mathbf{Q}}] &= \widehat{\boldsymbol{\mu}}^T\widehat{\boldsymbol{\mu}} - \frac{1}{M(M-1)}\sum_{m=1}^{M}\boldsymbol{\xi}_m^T\boldsymbol{\xi}_m + \frac{1}{M-1}\widehat{\boldsymbol{\mu}}^T\widehat{\boldsymbol{\mu}} \\ &= \frac{M}{M-1}\widehat{\boldsymbol{\mu}}^T\widehat{\boldsymbol{\mu}} - \frac{1}{M(M-1)}\sum_{m=1}^{M}\boldsymbol{\xi}_m^T\boldsymbol{\xi}_m.\end{aligned} \tag{6.12}$$

Substituting Eq. (6.10) into Eq. (6.12), the computation still goes on, and

$$\widehat{\boldsymbol{\mu}}^T\widehat{\boldsymbol{\mu}} - \frac{1}{M}\mathrm{Tr}[\widehat{\mathbf{Q}}] = \frac{1}{M(M-1)}\sum_{m\neq l}\boldsymbol{\xi}_m^T\boldsymbol{\xi}_l. \tag{6.13}$$

This shows that the estimate (6.13) removes the quadratic terms $m = l$ from which the bias of the naive estimate (6.10) was induced. However, note that formulation (6.12) is more convenient to compute than (6.13), from an algorithmic perspective.

6.4.2.2 Noise Estimation

The estimation of σ_ε^2 is separated into two cases, the non-regularised and the regularised ones. A couple of notations is needed to make readable the equations. The residuals for the mth model are $\mathbf{r}_m = \mathbf{y} - \mathbf{P}_m\mathbf{y}$, where $\mathbf{P}_m = \mathbf{H}_m\mathbf{H}_m^\alpha$. Also, $\mathbf{b} = \mathbf{f} - \mathbb{E}[\hat{\mathbf{f}}] = (\mathbf{I} - \mathbb{E}[\mathbf{P}])\mathbf{f}$ is the vector of bias of the model predictions at the training points $\hat{\mathbf{f}}$, and $\mathbf{b}_{w,m} = \mathbf{f} - \mathbb{E}[\hat{\mathbf{f}} \mid \mathbf{W}_m] = (\mathbf{I} - \mathbf{P}_m)\mathbf{f}$ is the vector of conditional bias of the model predictions at the training points $\hat{\mathbf{f}}$ for the mth model knowing $\mathbf{W}_m$.

Let us first concentrate on the non-regularised case. Then, $\mathbf{P}_m$ is a projection matrix. A natural way to obtain an estimate for $\hat{\sigma}_\varepsilon^2$ is to start with the expectation of the Residual Sum of Squares (RSS) based on the averaged ensemble. However, mainly since the expectation of a projection matrix is not a projection matrix, it is preferred here to work with the RSS of each model. Using this, the expectation of the RSS for the mth ELM knowing input weights and biases is given by

$$\begin{aligned}\mathbb{E}\left[\mathbf{r}_m^T\mathbf{r}_m \,\middle|\, \mathbf{W}_m\right] &= \mathbb{E}\left[\mathbf{y}^T(\mathbf{I} - \mathbf{P}_m)\mathbf{y} \,\middle|\, \mathbf{W}_m\right] \\ &= \sigma_\varepsilon^2\,\mathrm{Tr}[\mathbf{I} - \mathbf{P}_m] + \mathbf{f}^T(\mathbf{I} - \mathbf{P}_m)\mathbf{f} \\ &= \sigma_\varepsilon^2(n - N) + \mathbf{f}^T\mathbf{b}_{w,m},\end{aligned} \tag{6.14}$$

and taking the expectation over the input weights and biases yields $\mathbb{E}[\mathbf{r}_m^T\mathbf{r}_m] = \sigma_\varepsilon^2(n - N) + \mathbf{f}^T\mathbf{b}$. This motivates the following noise estimate,

$$\hat{\sigma}_\varepsilon^2 = \frac{1}{M(n-N)}\sum_{m=1}^{M}\mathbf{r}_m^T\mathbf{r}_m, \tag{6.15}$$

which is the average of all MLR estimates of σ_ε^2. Its bias is directly obtained from the previous calculation, yielding

$$\mathrm{Bias}\left[\hat{\sigma}_\varepsilon^2\right] = \frac{1}{n-N}\mathbf{f}^T\mathbf{b} \geq 0\,, \tag{6.16}$$

where non-negativity results from the facts that a projection matrix is positive semidefinite and expcctation of a positive semidefinite matrix is positive semidefinite.

If regularised ELMs are used, $\mathbf{P}_m$ is no more a projection matrix, and the expected RSS for each ELM of the ensemble knowing $\mathbf{W}_m$ becomes

$$\begin{aligned}\mathbb{E}\left[\mathbf{r}_m^T\mathbf{r}_m \,\middle|\, \mathbf{W}_m\right] &= \mathbb{E}\left[\mathbf{y}^T(\mathbf{I}-\mathbf{P}_m)^T(\mathbf{I}-\mathbf{P}_m)\mathbf{y} \,\middle|\, \mathbf{W}_m\right]\\ &= \sigma_\varepsilon^2\,\mathrm{Tr}\left[(\mathbf{I}-\mathbf{P}_m)^T(\mathbf{I}-\mathbf{P}_m)\right] + \mathbf{f}^T(\mathbf{I}-\mathbf{P}_m)^T(\mathbf{I}-\mathbf{P}_m)\mathbf{f}\\ &= \sigma_\varepsilon^2\left(n - 2\mathrm{Tr}\left[\mathbf{P}_m\right] + \mathrm{Tr}\left[\mathbf{P}_m^2\right]\right) + \mathbf{b}_{w,m}^T\mathbf{b}_{w,m}\,.\end{aligned} \tag{6.17}$$

Analogously to what is done in [22], the *effective degrees of freedom for error* can be defined as $n-\gamma$, with $\gamma = 2\mathbb{E}\left[\mathrm{Tr}\left[\mathbf{P}\right]\right] - \mathbb{E}\left[\mathrm{Tr}\left[\mathbf{P}^2\right]\right]$. Expectation over input weights and biases of Eq. (6.17) gives $\mathbb{E}\left[\mathbf{r}_m^T\mathbf{r}_m\right] = \sigma_\varepsilon^2\,(n-\gamma) + \mathbb{E}[\mathbf{b}_w^T\mathbf{b}_w]$, hence,

$$\begin{aligned}\mathbb{E}\left[\frac{1}{M(n-\gamma)}\sum_{m=1}^{M}\mathbf{r}_m^T\mathbf{r}_m\right] &= \sigma_\varepsilon^2 + \frac{1}{n-\gamma}\mathbb{E}\left[\mathbf{b}_w^T\mathbf{b}_w\right]\\ &= \sigma_\varepsilon^2 + \frac{1}{n-\gamma}\left(\mathbf{b}^T\mathbf{b} + \mathrm{Tr}\left[\mathrm{Var}\left[\mathbf{b}_w\right]\right]\right),\end{aligned}$$

which make appears the squared bias and the total variation of the conditional bias. This motivates the following noise estimate in the regularised case,

$$\hat{\sigma}_\varepsilon^2 = \frac{1}{M(n-\widehat{\gamma})}\sum_{m=1}^{M}\mathbf{r}_m^T\mathbf{r}_m\,,$$

with the estimated effective degrees of freedom for error,

$$\widehat{\gamma} = \frac{1}{M}\sum_{m=1}^{M}\left(2\mathrm{Tr}\left[\mathbf{P}_m\right] - \mathrm{Tr}\left[\mathbf{P}_m^2\right]\right). \tag{6.18}$$

It is easy to check that

$$\mathrm{Bias}\left[\hat{\sigma}_\varepsilon^2\right] = \mathbb{E}\left[\frac{1}{n-\widehat{\gamma}}\mathbf{b}_w^T\mathbf{b}_w\right] \geq 0\,. \tag{6.19}$$

Computationally, $\widehat{\gamma}$ can be efficiently calculated using the SVD of $\mathbf{H}_m$, presented in Sect. 3.1.3 of Chap. 3. Indeed, it can be shown [17, 23] that the trace of $\mathbf{P}_m$ and $\mathbf{P}_m^2$ are given by

$$\begin{aligned} \operatorname{Tr}\left[\mathbf{P}_m\right] &= \sum_{i=1}^{N} \frac{\lambda_{m,i}}{\lambda_{m,i}+\alpha}, \\ \operatorname{Tr}\left[\mathbf{P}_m^2\right] &= \sum_{i=1}^{N} \left(\frac{\lambda_{m,i}}{\lambda_{m,i}+\alpha}\right)^2, \end{aligned} \tag{6.20}$$

where $\lambda_{m,i}, i = 1, \ldots, N$ are the eigenvalues of $\mathbf{H}_m^T\mathbf{H}_m$. In particular, the substitution of Eqs. (6.20) in (6.18) and elementary manipulations allows to writes

$$\widehat{\gamma} = N - \frac{1}{M}\sum_{m=1}^{M}\sum_{i=1}^{N}\left(\frac{\alpha}{\lambda_{m,i}+\alpha}\right)^2$$

where $\sqrt{\lambda_{m,i}}, i = 1, \ldots, N$ are the singular values of $\mathbf{H}_m$. Note that this latter equation also shows the drop in the degrees of freedom lost due to regularisation, comparing to the non-regularised case.

6.4.2.3 Estimations of ELM Ensemble Variance

As a first guess, the following *Naive Homoskedastic* (NHo) estimate of the variance of $\hat{f}(\mathbf{x}_0)$ is proposed,

$$\hat{\sigma}^2_{\text{NHo}} = \hat{\sigma}^2_{\varepsilon}\,\widehat{\boldsymbol{\mu}}^T\widehat{\boldsymbol{\mu}} + \frac{1}{M}\hat{\sigma}^2_{\hat{f}}.$$

This naive estimate directly uses Eq. (6.10) to approximate $\boldsymbol{\mu}^T\boldsymbol{\mu}$ without considering its bias. However, a *Bias-Reduced* (BR) estimate is obtained by estimating $\boldsymbol{\mu}^T\boldsymbol{\mu}$ by Eq. (6.13), yielding

$$\hat{\sigma}^2_{\text{BR}} = \hat{\sigma}^2_{\varepsilon}\left(\widehat{\boldsymbol{\mu}}^T\widehat{\boldsymbol{\mu}} - \frac{1}{M}\text{Tr}[\widehat{\mathbf{Q}}]\right) + \frac{1}{M}\hat{\sigma}^2_{\hat{f}}.$$

Using the covariance definition, it is easy to see that

$$\text{Bias}\left[\hat{\sigma}^2_{\text{BR}}\right] = \text{Bias}\left[\hat{\sigma}^2_{\varepsilon}\right]\boldsymbol{\mu}^T\boldsymbol{\mu} + \text{Cov}\left[\hat{\sigma}^2_{\varepsilon},\ \widehat{\boldsymbol{\mu}}^T\widehat{\boldsymbol{\mu}} - \frac{1}{M}\text{Tr}[\widehat{\mathbf{Q}}]\right], \tag{6.21}$$

where the bias of $\hat{\sigma}^2_{\varepsilon}$ is (6.16) or (6.19). Note that in both cases, while the first term of the RHS of Eq. (6.21) is always non-negative, the estimates of σ^2_{ε} and $\boldsymbol{\mu}^T\boldsymbol{\mu}$ could be correlated, introducing the second term. However, this supplementary bias could be negative, potentially compensating for the first term. Note that its magnitude is bounded by

$$\left| \mathrm{Cov}\left[\hat{\sigma}_\varepsilon^2,\ \widehat{\boldsymbol{\mu}}^T\widehat{\boldsymbol{\mu}} - \frac{1}{M}\mathrm{Tr}[\widehat{\mathbf{Q}}]\right] \right| \leq \frac{2}{M(n-N)}\sqrt{\mathbf{f}^T\mathrm{Var}\left[\mathbf{b}_w\right]\mathbf{f}\ \cdot\ \boldsymbol{\mu}^T\mathbf{C}\boldsymbol{\mu}}, \tag{6.22}$$

in the non-regularised case—see calculation in the next paragraph—showing that this covariance term vanishes with large M. Remark also that the bias of $\hat{\sigma}^2_{\mathrm{NHo}}$ has an additional non-negative term, $(1/M)\mathrm{Tr}\left[\mathbf{C}\right]\mathbb{E}\left[\hat{\sigma}_\varepsilon^2\right]$, which disappears in (6.21) thanks to the unbiasedness of $\widehat{\boldsymbol{\mu}}^T\widehat{\boldsymbol{\mu}} - (1/M)\mathrm{Tr}[\widehat{\mathbf{Q}}]$.

A detailed derivation of the bound (6.22) is provided here. First, note that for all $m, l, k = 1, \ldots M$,

$$\begin{aligned}
\mathrm{Cov}\left[\mathbf{r}_m^T\mathbf{r}_m,\ \boldsymbol{\xi}_k^T\boldsymbol{\xi}_l\right] &= \mathrm{Cov}\left[\mathbb{E}\left[\mathbf{r}_m^T\mathbf{r}_m \,\middle|\, \mathbf{W}_m\right],\ \mathbb{E}\left[\boldsymbol{\xi}_k^T\boldsymbol{\xi}_l \,\middle|\, \mathbf{W}_l, \mathbf{W}_k\right]\right] \\
&\quad + \mathbb{E}\left[\mathrm{Cov}\left[\mathbf{r}_m^T\mathbf{r}_m,\ \boldsymbol{\xi}_k^T\boldsymbol{\xi}_l \,\middle|\, \mathbf{W}_m, \mathbf{W}_l, \mathbf{W}_k\right]\right] \\
&= \mathrm{Cov}\left[\mathbf{f}^T\mathbf{b}_{w,m},\ \boldsymbol{\xi}_k^T\boldsymbol{\xi}_l\right],
\end{aligned} \tag{6.23}$$

where the first equality uses the law of total covariance, and the second equality uses Eq. (6.14), the fact that expectation has no effect on a constant and that covariance between a random variable and a constant is null. Also, using the covariance definition and the independence of weights between models, one has for all $k \neq l$,

$$\begin{aligned}
\mathrm{Cov}\left[\mathbf{f}^T\mathbf{b}_{w,k},\ \boldsymbol{\xi}_k^T\boldsymbol{\xi}_l\right] &= \mathbb{E}\left[\mathbf{f}^T\mathbf{b}_{w,k}\boldsymbol{\xi}_k^T\boldsymbol{\xi}_l\right] - \mathbb{E}\left[\mathbf{f}^T\mathbf{b}_{w,k}\right]\mathbb{E}\left[\boldsymbol{\xi}_k^T\boldsymbol{\xi}_l\right] \\
&= \mathbb{E}\left[\mathbf{f}^T\mathbf{b}_{w,k}\boldsymbol{\xi}_k^T\right]\boldsymbol{\mu} - \mathbb{E}\left[\mathbf{f}^T\mathbf{b}_{w,k}\right]\mathbb{E}\left[\boldsymbol{\xi}_k^T\right]\boldsymbol{\mu} \\
&= \mathrm{Cov}\left[\mathbf{f}^T\mathbf{b}_{w,k},\ \boldsymbol{\xi}_k^T\boldsymbol{\mu}\right].
\end{aligned} \tag{6.24}$$

Then, using Eqs. (6.13), (6.15), (6.23) and (6.24), one obtains

$$\begin{aligned}
\mathrm{Cov}\left[\hat{\sigma}_\varepsilon^2,\ \widehat{\boldsymbol{\mu}}^T\widehat{\boldsymbol{\mu}} - \frac{1}{M}\mathrm{Tr}[\widehat{\mathbf{Q}}]\right] &= \frac{1}{M^2(M-1)(n-N)}\sum_{m=1}^{M}\sum_{k\neq l}\mathrm{Cov}\left[\mathbf{r}_m^T\mathbf{r}_m,\ \boldsymbol{\xi}_k^T\boldsymbol{\xi}_l\right] \\
&= \frac{1}{M^2(M-1)(n-N)}\sum_{m=1}^{M}\sum_{k\neq l}\mathrm{Cov}\left[\mathbf{f}^T\mathbf{b}_{w,m},\ \boldsymbol{\xi}_k^T\boldsymbol{\xi}_l\right] \\
&= \frac{2}{M^2(M-1)(n-N)}\sum_{k\neq l}\mathrm{Cov}\left[\mathbf{f}^T\mathbf{b}_{w,k},\ \boldsymbol{\xi}_k^T\boldsymbol{\xi}_l\right] \\
&= \frac{2}{M^2(M-1)(n-N)}\sum_{k\neq l}\mathrm{Cov}\left[\mathbf{f}^T\mathbf{b}_{w,k},\ \boldsymbol{\xi}_k^T\boldsymbol{\mu}\right] \\
&= \frac{2}{M(n-N)}\mathrm{Cov}\left[\mathbf{f}^T\mathbf{b}_w,\ \boldsymbol{\xi}^T\boldsymbol{\mu}\right],
\end{aligned}$$

where covariances vanish for $m \neq k, l$ in the third equality. By using the Cauchy-Schwarz inequality on the last equation, one gets

$$\left| \text{Cov}\left[\hat{\sigma}_\varepsilon^2,\ \widehat{\boldsymbol{\mu}}^T\widehat{\boldsymbol{\mu}} - \frac{1}{M}\text{Tr}[\widehat{\mathbf{Q}}]\right] \right| \leq \frac{2}{M(n-N)}\sqrt{\text{Var}\left[\mathbf{f}^T\mathbf{b}_w\right]\ \text{Var}\left[\boldsymbol{\xi}^T\boldsymbol{\mu}\right]}$$

from which Eq. (6.22) is obtained.

6.4.3 Estimation Under Independence and Heteroskedastic Assumptions

Suppose the noise is independent but has variance which has a dependence of unknown form on $\mathbf{x}$. Then $S = \boldsymbol{\mu}^T\boldsymbol{\Sigma}\boldsymbol{\mu}$ has to be estimated considering the noise covariance matrix $\boldsymbol{\Sigma}$ as diagonal. To this aim, it could be possible to reuse estimates from MLR. However, several estimates are based on the evaluation of the covariance matrix $\mathbf{H}_m^\alpha\boldsymbol{\Sigma}_m\mathbf{H}_m^{\alpha T}$ of the output weights $\widehat{\boldsymbol{\beta}}_m$. In this thesis, the modified Heteroskedastic-Consistent Covariance Matrix Estimator (HCCME) obtained from the (ordinary) Jackknife [24]—noted HC_3 and extended to the ridge regression case [25]—is used,

$$\text{HC}_3 = \mathbf{H}_m^\alpha\widehat{\boldsymbol{\Sigma}}_m\mathbf{H}_m^{\alpha T}, \quad \text{with} \quad \widehat{\boldsymbol{\Sigma}}_m = \frac{n-1}{n}\left[\widehat{\boldsymbol{\Omega}}_m - \frac{1}{n}\tilde{\mathbf{r}}_m\tilde{\mathbf{r}}_m^T\right], \tag{6.25}$$

where $\tilde{\mathbf{r}}_m$ is the vector defined element-wise by $\tilde{\mathbf{r}}_{m,i} = \mathbf{r}_{m,i}/(1 - p_{m,i})$, $p_{m,i}$ is the ith diagonal element of $\mathbf{P}_m$ and $\widehat{\boldsymbol{\Omega}}_m$ is the diagonal matrix with the ith diagonal element equal to $\tilde{\mathbf{r}}_{m,i}^2$. This estimate is still valid for the non-regularised case, for which—under some technical assumptions—it is consistent [24, 26], while $\widehat{\boldsymbol{\Sigma}}_m$ is an inconsistent estimator of $\boldsymbol{\Sigma}$. Nevertheless, other HCCME estimates could be used, such as HC_0 [26], HC_1 got from the weighted Jackknife [27], HC_2 proposed in [28] or HC_4 proposed more recently in [29]. The HC notation follows what can be found in [13], which provides useful insights into this kind of estimators. Note that for sufficiently large n, HC_3 is close to $\mathbf{H}_m^\alpha\widehat{\boldsymbol{\Omega}}_m\mathbf{H}_m^{\alpha T}$, which corresponds to the estimate used in [9] to build prediction intervals for large amounts of data, assuming fixed input weights.

If a unique ELM model is performed, the use of the HCCME is straightforward. Nonetheless, as one attempts to consider the input weight variability through ELM replications, the HCCME is applied in three different ways. Suppose first that $\boldsymbol{\Sigma}$ is known and write the estimate

$$\widehat{\boldsymbol{\mu}}^T\boldsymbol{\Sigma}\widehat{\boldsymbol{\mu}} = \frac{1}{M^2}\sum_{m,l=1}^{M}\boldsymbol{\xi}_m^T\boldsymbol{\Sigma}\boldsymbol{\xi}_l = \frac{1}{M^2}\sum_{m,l=1}^{M}\mathbf{h}_m^T\mathbf{H}_m^\alpha\boldsymbol{\Sigma}\mathbf{H}_l^{\alpha T}\mathbf{h}_l. \tag{6.26}$$

Inspecting Eqs. (6.25) and (6.26), a first natural suggestion is to estimate $\boldsymbol{\mu}^T \boldsymbol{\Sigma} \boldsymbol{\mu}$ with

$$\widehat{S}_1 = \frac{1}{M} \sum_{m=1}^{M} \boldsymbol{\xi}_m^T \widehat{\boldsymbol{\Sigma}}_m \boldsymbol{\xi}_m.$$

Note that $\widehat{S}_1$ estimates the covariance matrix of the output weights with the HCCME for each of the M random feature spaces. Although it has the advantage of reusing the HCCME in its original formulation, the quadratic forms in random vectors depending on input weight may produce an additional bias. Another estimator is obtained by naively evaluating all cross-terms of Eq. (6.26),

$$\widehat{S}_{\text{NHe}} = \frac{1}{M^2} \sum_{m,l=1}^{M} \boldsymbol{\xi}_m^T \widehat{\boldsymbol{\Sigma}}_m \boldsymbol{\xi}_l.$$

Note that this is equivalent to estimate $\boldsymbol{\Sigma}$ with $(\widehat{\boldsymbol{\Sigma}}_m + \widehat{\boldsymbol{\Sigma}}_l)/2$ in Eq. (6.26), using the symmetry of $\widehat{\boldsymbol{\Sigma}}_l$ and the transpose operator on scalars. However, terms for which $m = l$ may produce additional biases, similarly to what was shown for the homoskedastic estimate of Eq. (6.10), which motivates

$$\widehat{S}_2 = \frac{1}{M(M-1)} \sum_{m \neq l} \boldsymbol{\xi}_m^T \widehat{\boldsymbol{\Sigma}}_m \boldsymbol{\xi}_l.$$

Analogously to the homoskedastic case—see Eq. (6.13)—the terms corresponding to $m = l$ are not taken into account in $\widehat{S}_2$, avoiding the introduction of potential biases from quadratic forms. Remark also that

$$\widehat{S}_2 = \frac{1}{M-1} \left[M \widehat{\boldsymbol{\nu}}^T \widehat{\boldsymbol{\mu}} - \widehat{S}_1 \right], \quad \text{where} \quad \widehat{\boldsymbol{\nu}} = \frac{1}{M} \sum_{m=1}^{M} \widehat{\boldsymbol{\Sigma}}_m \boldsymbol{\xi}_m,$$

which is algorithmically more convenient to compute. Also, $\widehat{S}_{\text{NHe}} = \widehat{\boldsymbol{\nu}}^T \widehat{\boldsymbol{\mu}}$, which is more efficient than evaluating all pairs of models. The estimates $\widehat{S}_{\text{NHe}}$ and $\widehat{S}_2$ have some similarities with $\widehat{\boldsymbol{\mu}}^T \widehat{\boldsymbol{\mu}}$ and $\widehat{\boldsymbol{\mu}}^T \widehat{\boldsymbol{\mu}} - (1/M)\text{Tr}\,[\mathbf{C}]$ as estimated in the homoskedastic case; see Sect. 6.4.2.1. However, $\boldsymbol{\xi}_m$ still interacts with the covariance matrix estimate within $\widehat{\boldsymbol{\nu}}$. This motivates a third estimate,

$$\widehat{S}_3 = \frac{(M-3)!}{M!} \sum_{(m,l,k) \in A_M^3} \boldsymbol{\xi}_m^T \widehat{\boldsymbol{\Sigma}}_k \boldsymbol{\xi}_l,$$

where A_M^3 is the set of 3-permutations of M. Looking at Eq. (6.26), $\widehat{S}_3$ can also be obtained by replacing $\boldsymbol{\Sigma}$ by a single estimate consisting of the average of the estimate of each model, where terms corresponding to $m = l$, $m = k$, or $l = k$ are ignored to avoid additional biases. To compute efficiently $\widehat{S}_3$, it can be rewritten as

$$\widehat{S}_3 = \frac{1}{(M-1)(M-2)} \left[M^2 \widehat{\mu}^T \widehat{\mathbf{U}} \widehat{\mu} - M \widehat{V} - 2(M-1)\widehat{S}_2 \right],$$

with

$$\widehat{\mathbf{U}} = \frac{1}{M} \sum_{m=1}^{M} \widehat{\mathbf{\Sigma}}_m \quad \text{and} \quad \widehat{V} = \frac{1}{M} \sum_{m=1}^{M} \boldsymbol{\xi}_m^T \widehat{\mathbf{U}} \boldsymbol{\xi}_m.$$

Finally, the proposed heteroskedastic estimates of ELM ensemble variance, noted $\hat{\sigma}^2_{S1}$, $\hat{\sigma}^2_{S2}$, $\hat{\sigma}^2_{S3}$ and $\hat{\sigma}^2_{\text{NHe}}$, are given by respectively adding $(1/M)\hat{\sigma}^2_{\hat{f}}$ to $\widehat{S}_1$, $\widehat{S}_2$, $\widehat{S}_3$ and $\widehat{S}_{\text{NHe}}$.

To increase computation speed, approximated versions of $\widehat{S}_1$, $\widehat{S}_2$, $\widehat{S}_3$, and $\widehat{S}_{\text{NHe}}$ can be obtained by replacing $\widehat{\mathbf{\Sigma}}_m$ with $\widehat{\mathbf{\Omega}}_m$ in the above reasoning. As a matter of fact, these two matrices are very close for sufficiently large n, but $\widehat{\mathbf{\Omega}}_m$ is a diagonal matrix, while $\widehat{\mathbf{\Sigma}}_m$ is a full matrix.

6.5 Synthetic Experiments

This section discusses the results obtained over different experimental settings. First, a simple non-regularised homoskedastic one-dimensional experiment is conducted. The variance estimate is thoroughly examined and assessed with quantitative measures and visualisations. Subsequently, the results are generalised to multi-dimensional settings with homoskedastic or heteroskedastic noise, both for the regularised and non-regularised cases. Finally, CI estimation is discussed. All the experiments presented will adopt the sigmoid as an activation function, while input weights and biases will always be drawn uniformly between -1 and 1. All computations are done with the provided Python library; see the software availability in Sect. 1.5.

6.5.1 *One-Dimensional Case*

In order to assess operationally the estimates proposed in Sect. 6.4, a simple one-dimensional simulated case study of $n = 60$ training points is firstly proposed. A trapeze shape PDF defined by

$$\rho(x) = -\frac{x}{4\pi^2} + \frac{3}{4\pi}, \quad \text{if } x \in [0, 2\pi],$$

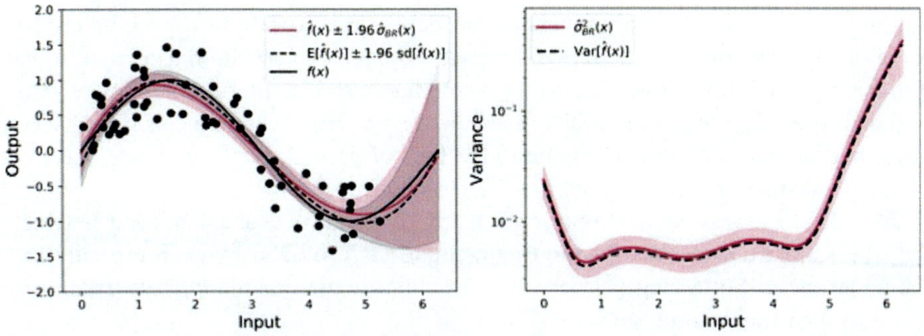

Fig. 6.1 **One-dimensional synthetic experiment**: (left) In magenta, a single estimation of $\hat{f}(x)$ with $M = 10$ and its estimated ± 1.96 standard-error bands based on $\hat{\sigma}^2_{\text{BR}}$. In black dashed line, the mean of 10'000 ensembles for $M = 10$, with ± 1.96 standard-error bands. The true $f(x)$ is displayed in full black line; (right) In magenta, the mean of $\hat{\sigma}^2_{\text{BR}}$ for $M = 10$ with ± 1.96 standard-error bands, based on 1'000 replications of the experiment. In black dashed line, the variance computed from the 10'000 ensembles, considered as ground truth. Note the logarithmic scale on the y-axis

$\rho(x) = 0$ otherwise, is used to draw the input, i.e. the number of data decreases as x increases. Outputs are generated according to

$$y = \sin(x) + \varepsilon,$$

where ε is an independent uniform noise $]-\sqrt{0.3}, \sqrt{0.3}[$ of constant variance $\sigma^2_\varepsilon = 0.1$.

ELM ensembles are trained with $M = 5, 10, 20, 100$, allowing variance estimation. This experience is repeated 1'000 times. Each time, new outputs and new weights are drawn, but inputs are fixed. In order to avoid variability induced by the hyper-parameter selection, a fixed number of neurons $N = 4$ was chosen by a 5-fold cross-validation process repeated 5 times on 1'000 data set generations. An example of one prediction is displayed in Fig. 6.1 (left) for $M = 10$. Estimation of pointwise standard-error bands based on $\pm 1.96\,\hat{\sigma}_{\text{BR}}$ is also reported.

Although simulated data sets are produced by user-controlled processes, the true value of the variance of $\hat{f}(x)$ remains unknown. In order to evaluate the estimate, $10'000$ ensembles with $M = 5, 10, 20, 100$ and $N = 4$ are trained with new outputs. The empirical mean and standard deviation of the $10'000$ ensembles is relatively close to respectively $\mathbb{E}[\hat{f}(x)]$ and $\text{sd}[\hat{f}(x)] = (\text{Var}[\hat{f}(x)])^{1/2}$, and is reported as such for $M = 10$ in Fig. 6.1. In particular, the empirical variance of the $10'000$ ensembles will provide a reliable baseline for the variance estimation assessment and will be referred to as the *ground truth variance*.

Figure 6.1 (right) shows the mean of $\hat{\sigma}^2_{\text{BR}}$ across the $1'000$ experiments with ± 1.96 standard-error band. Compared with the ground truth variance, the proposed estimate recovers effectively—on average—the variance from the $10'000$ simulations baseline. The increasing variance in the borders due to the side effect of the modelling is fairly replicated. The uncertainty due to the trapezoidal shape of the input data distribution is also captured. Qualitatively, all aspects of the expected variance behaviour are globally reproduced. The naive estimate $\hat{\sigma}^2_{\text{NHo}}$ gives very similar results in one dimension and is not shown. The improvement due to considering the bias of $\widehat{\boldsymbol{\mu}}^T\widehat{\boldsymbol{\mu}}$ will be more relevant in the multi-dimensional case. However, one can still observe a residual bias for $\hat{\sigma}^2_{\text{BR}}$, partially due to the bias of $\hat{\sigma}^2_\varepsilon$ and the dependence between $\hat{\sigma}^2_\varepsilon$ and the estimate of $\boldsymbol{\mu}^T\boldsymbol{\mu}$—see Eqs. (6.16) and (6.21).

In order to assess each estimation quantitatively, a measure is needed between the true standard error of $\hat{f}(\mathbf{x})$ and its estimations provided by each repetition of the experiment. Following [2], let us look at the median of the kth standard error estimate over the training set,

$$se_k = \underset{1\leq i\leq n}{\text{median}}\left(\hat{\sigma}_k(\mathbf{x}_i)\right),$$

and the absolute error of the kth standard error estimate over the training set defined by

$$e_k = \underset{1\leq i\leq n}{\text{median}}\,|\hat{\sigma}_k(\mathbf{x}_i) - \sigma(\mathbf{x}_i)|,$$

where $\hat{\sigma}^2_k(\mathbf{x}_i)$ is an estimate of $\sigma^2(\mathbf{x}_i) = \text{Var}[\hat{f}(\mathbf{x}_i)]$, for $k = 1, \ldots, 1'000$. Also, the relative error re_k of the kth standard error estimate over the training set is defined by

$$re_k = \underset{1\leq i\leq n}{\text{median}}\,\frac{|\hat{\sigma}_k(\mathbf{x}_i) - \sigma(\mathbf{x}_i)|}{\sigma(\mathbf{x}_i)},$$

for $k = 1, \ldots, 1'000$. Similar measures are defined on a random testing set of $1'000$ points. In order to compute these quantities, $\sigma(\mathbf{x}_i)$ is replaced by the ground truth standard deviation.

The means and standard deviations of se_k, e_k and re_k over the 1'000 experiment repetitions are presented in Table 6.1. For the training set, the median of the ground truth standard error is recovered by the median of $\hat{\sigma}_{\text{BR}}$, judging through the se_k measure. Moreover, the mean and standard deviation of e_k appear relatively small. The relative errors allow a better interpretation by comparing point-wise the absolute error with the true standard error. For instance, for $M = 10$, the mean of re_k shows that—on average—the median error at training points represents 6.2% of the true standard error. The results on the 1'000 testing points are similar, which shows that the estimation is good both at testing and training points. Even for $M = 5$, all error measures are quite satisfactory, as well as their standard deviations. These results are also compared with $\hat{\sigma}^2_{S3}$. As expected for an homoskedastic data set, estimate $\hat{\sigma}^2_{\text{BR}}$ based on homoskedastic assumption always results in better performance than $\hat{\sigma}^2_{S3}$ which is based on heteroskedastic assumption.

The same experiment is done with $n = 100$ and $N = 5$. The results, reported in Table 6.1, show that all results are improved when increasing the number of data points, as expected.

Table 6.1 Results of the one-dimensional synthetic experiment. Mean (standard deviation) of se_k, e_k and re_k

		$n = 60$			$n = 100$		
$M = 5$		BR	S3	Grnd tr.	BR	S3	Grnd tr.
Training set	se_k	0.077 (0.006)	0.080 (0.007)	0.075	0.058 (0.003)	0.060 (0.004)	0.058
	e_k	0.005 (0.004)	0.009 (0.005)	—	0.003 (0.002)	0.005 (0.002)	—
	re_k	0.066 (0.052)	0.113 (0.065)	—	0.046 (0.034)	0.077 (0.039)	—
Testing set	se_k	0.077 (0.006)	0.081 (0.007)	0.075	0.060 (0.003)	0.062 (0.004)	0.059
	e_k	0.005 (0.004)	0.009 (0.005)	—	0.003 (0.002)	0.005 (0.002)	—
	re_k	0.068 (0.053)	0.115 (0.066)	—	0.044 (0.033)	0.080 (0.039)	—
$M = 10$		BR	S3	Grnd tr.	BR	S3	Grnd tr.
Training set	se_k	0.076 (0.005)	0.079 (0.006)	0.074	0.058 (0.003)	0.060 (0.004)	0.057
	e_k	0.005 (0.003)	0.008 (0.004)	—	0.003 (0.002)	0.004 (0.002)	—
	re_k	0.062 (0.046)	0.109 (0.057)	—	0.044 (0.032)	0.075 (0.037)	—
Testing set	se_k	0.076 (0.005)	0.080 (0.006)	0.074	0.060 (0.003)	0.062 (0.004)	0.059
	e_k	0.005 (0.003)	0.009 (0.004)	—	0.003 (0.002)	0.005 (0.002)	—
	re_k	0.062 (0.046)	0.111 (0.056)	—	0.043 (0.032)	0.078 (0.037)	—
$M = 20$		BR	S3	Grnd tr.	BR	S3	Grnd tr.
Training set	se_k	0.076 (0.005)	0.079 (0.006)	0.074	0.058 (0.003)	0.060 (0.003)	0.057
	e_k	0.005 (0.003)	0.008 (0.004)	—	0.003 (0.002)	0.004 (0.002)	—
	re_k	0.061 (0.045)	0.109 (0.056)	—	0.042 (0.031)	0.074 (0.036)	—
Testing set	se_k	0.076 (0.005)	0.080 (0.006)	0.074	0.059 (0.003)	0.061 (0.004)	0.059
	e_k	0.005 (0.003)	0.008 (0.004)	—	0.003 (0.002)	0.005 (0.002)	—
	re_k	0.061 (0.044)	0.110 (0.056)	—	0.042 (0.031)	0.077 (0.037)	—
$M = 100$		BR	S3	Grnd tr.	BR	S3	Grnd tr.
Training set	se_k	0.076 (0.005)	0.079 (0.006)	0.074	0.058 (0.003)	0.060 (0.003)	0.057
	e_k	0.004 (0.003)	0.008 (0.004)	—	0.002 (0.002)	0.004 (0.002)	—
	re_k	0.060 (0.045)	0.109 (0.055)	—	0.041 (0.031)	0.073 (0.036)	—
Testing set	se_k	0.076 (0.005)	0.080 (0.006)	0.074	0.059 (0.003)	0.061 (0.004)	0.059
	e_k	0.004 (0.003)	0.008 (0.004)	—	0.002 (0.002)	0.005 (0.002)	—
	re_k	0.060 (0.044)	0.110 (0.055)	—	0.041 (0.031)	0.077 (0.036)	—

6.5.2 Multi-dimensional Case

A multi-dimensional example is now investigated. Specifically, the synthetic data set described by Friedman in [30] is considered, with fixed inputs $\mathbf{x} = (x_1, x_2, x_3, x_4, x_5)$ drawn independently from uniform distribution on the interval [0, 1] and outputs generated with an independent homoskedastic Gaussian noise according to

$$y(\mathbf{x}) = 10\sin(\pi x_1 x_2) + 20(x_3 - 0.5)^2 + 10x_4 + 5x_5 + \varepsilon, \tag{6.27}$$

with noise variance $\sigma_\varepsilon^2 = 0.5$. A number of $n = 500$ training points are drawn. The number of neurons is chosen by a similar cross-validation process as described above for the one-dimensional case and fixed to $N = 91$. Ensembles are fitted and homoskedastic estimates $\hat{\sigma}^2_{\text{BR}}$ and $\hat{\sigma}^2_{\text{NHo}}$ are computed with $M = 5, 10, 20, 100$. This is repeated 1′000 times, while ground truth mean and variance are computed based on 10’000 ensembles, as in the previous experiment. The same experiment is conducted with ensembles of regularised ELM. The Tikhonov factor is selected with the help of Generalised Cross-Validation (GCV) [17, 23] repeated on 1’000 data set generations and set to $\alpha = 6 \cdot 10^{-6}$. The regularised version of $\hat{\sigma}^2_{\text{BR}}$ and $\hat{\sigma}^2_{\text{NHo}}$ are computed.

Results of the regularised and non-regularised versions of the experiment are reported in Table 6.2. For both, the training se_k tends to slightly overestimate the true standard deviation median over the training points. The testing se_k has analogous behaviour. Although the testing se_k tends globally to be greater than the training se_k, the testing e_k and re_k are similar to the training e_k and re_k. It suggests that although the prediction is more uncertain at testing points, the variance estimation works at testing points as well as at the training points, as in the one-dimensional experiment. Note also that the true standard deviation median decreases as M increases, as suggested by Eq. (6.8). For the non-regularised case, the bias-reduced estimate $\hat{\sigma}^2_{\text{BR}}$ is systematically better than the $\hat{\sigma}^2_{\text{NHo}}$ estimate. Recalling that $\hat{\sigma}^2_{\text{BR}}$ reduce the bias by a quantity inversely proportional to M—see Sect. 6.4.2—one observes that for e_k and re_k the improvement over $\hat{\sigma}^2_{\text{NHo}}$ is decreasing with M. The regularisation mechanism increases the model's bias while its variance decreases, which explains the decreasing of the true standard deviation median for a given M from the non-regularised to the regularised case. Moreover, for the regularised case, as the bias of the variance estimation depends directly on the conditional bias of the model, this could explain that the regularised experiment yields slightly weaker results in terms of e_k and re_k. However, observe that $\hat{\sigma}^2_{\text{BR}}$ is still better than $\hat{\sigma}^2_{\text{NHo}}$.

The same experiment is conducted with a non-constant noise variance to illustrate the heteroskedastic case. The Gaussian noise $\varepsilon(\mathbf{x})$ is now depending on the inputs variables through its variance by

$$\sigma_\varepsilon^2(\mathbf{x}) = 0.5 + 2\sin^2(\pi||\mathbf{x}||_\infty),$$

Table 6.2 Results of the multi-dimensional synthetic experiment with homoskedastic noise. Mean (standard deviation) of se_k, e_k and re_k

		Non-regularised			Regularised		
$M = 5$		BR	NHo	Grnd tr.	BR	NHo	Grnd tr.
Training set	se_k	0.273 (0.009)	0.279 (0.009)	0.255	0.250 (0.009)	0.252 (0.009)	0.222
	e_k	0.018 (0.008)	0.023 (0.009)	—	0.025 (0.009)	0.027 (0.009)	—
	re_k	0.070 (0.032)	0.089 (0.037)	—	0.112 (0.039)	0.121 (0.039)	—
Testing set	se_k	0.282 (0.010)	0.289 (0.010)	0.264	0.253 (0.009)	0.255 (0.009)	0.228
	e_k	0.018 (0.008)	0.024 (0.009)	—	0.025 (0.009)	0.027 (0.009)	—
	re_k	0.069 (0.030)	0.090 (0.034)	—	0.110 (0.039)	0.119 (0.039)	—
$M = 10$		BR	NHo	Grnd tr.	BR	NHo	Grnd tr.
Training set	se_k	0.266 (0.009)	0.269 (0.009)	0.247	0.243 (0.008)	0.244 (0.008)	0.214
	e_k	0.019 (0.008)	0.022 (0.008)	—	0.028 (0.008)	0.029 (0.008)	—
	re_k	0.078 (0.033)	0.089 (0.034)	—	0.128 (0.035)	0.132 (0.035)	—
Testing set	se_k	0.274 (0.009)	0.277 (0.009)	0.255	0.246 (0.008)	0.247 (0.008)	0.219
	e_k	0.019 (0.008)	0.023 (0.009)	—	0.028 (0.008)	0.029 (0.008)	—
	re_k	0.078 (0.033)	0.090 (0.034)	—	0.127 (0.035)	0.132 (0.035)	—
$M = 20$		BR	NHo	Grnd tr.	BR	NHo	Grnd tr.
Training set	se_k	0.263 (0.008)	0.264 (0.008)	0.243	0.239 (0.007)	0.240 (0.007)	0.210
	e_k	0.020 (0.008)	0.021 (0.008)	—	0.028 (0.007)	0.029 (0.007)	—
	re_k	0.082 (0.033)	0.087 (0.033)	—	0.134 (0.034)	0.137 (0.034)	—
Testing set	se_k	0.270 (0.008)	0.272 (0.009)	0.250	0.242 (0.007)	0.243 (0.007)	0.214
	e_k	0.020 (0.008)	0.022 (0.008)	—	0.028 (0.007)	0.029 (0.007)	—
	re_k	0.081 (0.033)	0.088 (0.033)	—	0.134 (0.034)	0.137 (0.034)	—
$M = 100$		BR	NHo	Grnd tr.	BR	NHo	Grnd tr.
Training set	se_k	0.259 (0.008)	0.260 (0.008)	0.239	0.236 (0.007)	0.236 (0.007)	0.207
	e_k	0.020 (0.008)	0.020 (0.008)	—	0.029 (0.007)	0.029 (0.007)	—
	re_k	0.084 (0.033)	0.085 (0.033)	—	0.139 (0.033)	0.139 (0.033)	—
Testing set	se_k	0.267 (0.008)	0.267 (0.008)	0.239	0.239 (0.007)	0.239 (0.007)	0.210
	e_k	0.021 (0.008)	0.021 (0.008)	—	0.029 (0.007)	0.029 (0.007)	—
	re_k	0.084 (0.033)	0.086 (0.033)	—	0.140 (0.033)	0.140 (0.033)	—

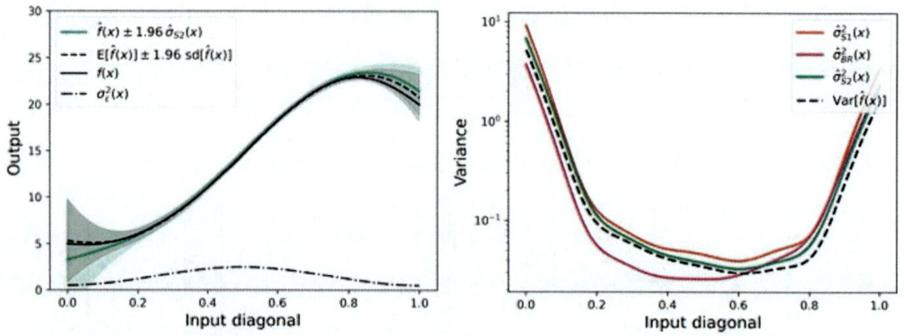

Fig. 6.2 Multi-dimensional heteroskedastic synthetic experiment : (left) In green, a single estimation of $\hat{f}(x)$ with $M = 5$ and its estimated ± 1.96 standard-error bands based on $\hat{\sigma}^2_{S2}$. In black dashed line, the mean of 10'000 ensembles for $M = 5$, with ± 1.96 standard-error bands. The true $f(x)$ is displayed in full black line and the noise variance in dash-dotted black line; (right) The averaged variance estimations for $\hat{\sigma}^2_{S2}$ in green, $\hat{\sigma}^2_{S1}$ in red and $\hat{\sigma}^2_{\text{BR}}$ in purple, based on 1'000 replications of the experiment. The black dashed line indicates the variance computed from the 10'000 ensembles, considered as ground truth. The diagonal distance from the origin on the x-axis is measured with the maximum norm. Note the logarithmic scale on the y-axis

where $||\cdot||_\infty$ denotes the maximum norm. The variance estimates $\hat{\sigma}^2_{S3}$, $\hat{\sigma}^2_{S2}$, $\hat{\sigma}^2_{\text{NHe}}$, $\hat{\sigma}^2_{S1}$ and $\hat{\sigma}^2_{\text{BR}}$ are computed in their approximated version $1'000$ times with $n = 1000$, $M = 5, 10, 20, 100$, and $N = 109$.

Although the results on the 5-dimensional hypercube input cannot be visualised, a small subset such as its diagonal can be plotted; see Fig. 6.2. On the left, prediction with $\pm 1.96\,\hat{\sigma}_{S2}$ is displayed for one experiment, for $M = 5$. The true noise variance $\sigma^2_\varepsilon(\mathbf{x})$ is also reported. On the right, averaged results for $\hat{\sigma}^2_{S2}$, $\hat{\sigma}^2_{S1}$ and $\hat{\sigma}^2_{\text{BR}}$ for the 1'000 experiments are shown. Results for $\hat{\sigma}^2_{S3}$ and σ^2_{NHe} are visually close to $\hat{\sigma}^2_{S2}$ and are not reported. Heteroskedastic estimates reproduce fairly well the behaviour of the true variance, but $\hat{\sigma}^2_{S2}$ shows a smaller bias than $\hat{\sigma}^2_{S1}$ along with the input diagonal. The homoskedastic estimates $\hat{\sigma}^2_{\text{BR}}$ fails to reproduce a coherent behaviour of the true variance, underestimating or overestimating it, depending on the location.

Quantitative results are shown in Table 6.3. Again, the true standard deviation median decreases when M increases, and the training (testing) se_k tends somewhat to overestimate the true training (testing) standard deviation median.

The homoskedastic estimate $\hat{\sigma}^2_{\text{BR}}$ no longer gives the best results because of the heteroskedastic nature of the data, which justify the use of heteroskedastic estimates. The estimate $\hat{\sigma}^2_{S1}$—which reuse HCCME in its original form—gives the worst results as suspected in Sect. 6.4.3. The naive heteroskedastic estimate $\hat{\sigma}^2_{\text{NHe}}$ gives in general reasonable results. However, the heteroskedastic estimates $\hat{\sigma}^2_{S2}$—which was developed based on the insight given by the homoskedastic case in Sect. 6.4.2—allows improving the results by a quantity decreasing with M, as expected. Finally, results from $\hat{\sigma}^2_{S3}$ are very close to $\hat{\sigma}^2_{S2}$.

Regularised versions of the heteroskedastic estimates were also investigated and $\hat{\sigma}^2_{S2}$ and $\hat{\sigma}^2_{S3}$ gave quite reasonable results too.

Table 6.3 Results of the multi-dimensional synthetic experiment with heteroskedastic noise. Mean (standard deviation) of se_k, e_k and re_k

$M = 5$		S3	S2	NHe	S1	BR	Grnd tr.
Training set	se_k	0.279 (0.008)	0.278 (0.007)	0.285 (0.008)	0.308 (0.008)	0.283 (0.007)	0.253
	e_k	0.027 (0.005)	0.027 (0.005)	0.031 (0.006)	0.052 (0.008)	0.046 (0.004)	—
	re_k	0.109 (0.019)	0.109 (0.019)	0.125 (0.024)	0.211 (0.032)	0.178 (0.015)	—
Testing set	se_k	0.302 (0.009)	0.302 (0.009)	0.309 (0.009)	0.335 (0.010)	0.294 (0.008)	0.275
	e_k	0.029 (0.006)	0.029 (0.006)	0.035 (0.007)	0.060 (0.009)	0.043 (0.004)	—
	re_k	0.111 (0.022)	0.111 (0.021)	0.131 (0.027)	0.230 (0.034)	0.157 (0.016)	—
$M = 10$		S3	S2	NHe	S1	BR	Grnd tr.
Training set	se_k	0.273 (0.007)	0.273 (0.007)	0.276 (0.007)	0.303 (0.008)	0.277 (0.007)	0.247
	e_k	0.027 (0.005)	0.027 (0.005)	0.029 (0.005)	0.053 (0.008)	0.047 (0.004)	—
	re_k	0.111 (0.020)	0.111 (0.020)	0.120 (0.022)	0.224 (0.032)	0.187 (0.015)	—
Testing set	se_k	0.296 (0.008)	0.296 (0.008)	0.299 (0.008)	0.330 (0.010)	0.288 (0.008)	0.267
	e_k	0.030 (0.006)	0.030 (0.006)	0.032 (0.006)	0.062 (0.009)	0.045 (0.004)	—
	re_k	0.115 (0.022)	0.114 (0.022)	0.126 (0.025)	0.246 (0.034)	0.168 (0.016)	—
$M = 20$		S3	S2	NHe	S1	BR	Grnd tr.
Training set	se_k	0.270 (0.007)	0.270 (0.007)	0.272 (0.007)	0.301 (0.008)	0.274 (0.007)	0.244
	e_k	0.027 (0.005)	0.027 (0.005)	0.028 (0.005)	0.054 (0.007)	0.048 (0.004)	—
	re_k	0.113 (0.020)	0.113 (0.020)	0.118 (0.021)	0.230 (0.031)	0.192 (0.015)	—
Testing set	se_k	0.292 (0.008)	0.292 (0.008)	0.294 (0.008)	0.327 (0.009)	0.284 (0.007)	0.263
	e_k	0.030 (0.006)	0.030 (0.006)	0.031 (0.006)	0.063 (0.009)	0.046 (0.004)	—
	re_k	0.117 (0.023)	0.117 (0.023)	0.123 (0.024)	0.254 (0.033)	0.173 (0.016)	—
$M = 100$		S3	S2	NHe	S1	BR	Grnd tr.
Training set	se_k	0.268 (0.007)	0.268 (0.007)	0.268 (0.007)	0.299 (0.008)	0.272 (0.007)	0.242
	e_k	0.027 (0.005)	0.027 (0.005)	0.028 (0.005)	0.055 (0.007)	0.049 (0.004)	—
	re_k	0.115 (0.020)	0.115 (0.020)	0.116 (0.020)	0.236 (0.031)	0.196 (0.015)	—
Testing set	se_k	0.290 (0.008)	0.290 (0.008)	0.290 (0.008)	0.325 (0.009)	0.282 (0.007)	0.261
	e_k	0.030 (0.006)	0.030 (0.006)	0.030 (0.006)	0.064 (0.008)	0.046 (0.004)	—
	re_k	0.120 (0.023)	0.120 (0.023)	0.121 (0.023)	0.260 (0.033)	0.178 (0.017)	—

6.5.3 Towards Confidence Intervals

Although visually it is tempting to say so, there is so far no guarantee that $\hat{f}(x) \pm 1.96\,\hat{\sigma}_{\mathrm{BR}}(x)$ in Fig. 6.1 (left) or $\hat{f}(x) \pm 1.96\,\hat{\sigma}_{S2}(x)$ in Fig. 6.2 (left) define a 95% CI for $f(x)$. Let us investigate the possibility to build (approximate) CI in particular cases. The distribution of

$$g(\mathbf{x}) = \frac{\hat{f}(\mathbf{x}) - f(\mathbf{x})}{\mathrm{sd}[\hat{f}(\mathbf{x})]},$$

is unknown. However, for our simulated case studies, remark that it is very close to a Gaussian distribution. KDE based on the 10'000 replications done in previous sections are shown in Fig. 6.3 at some example points at which $\hat{f}(\mathbf{x})$ exhibits significant bias. Figure 6.3 (left) displays the distribution of $g(x_0)$ at $x_0 = \pi/4$ for the one-dimensional case. Figure 6.3 (middle) visualises the distribution of $g(\mathbf{x}_0)$ at $\mathbf{x}_0 = 1/2 \cdot (1, 1, 1, 1, 1)$ for the Friedman data set with homoskedastic noise, which corresponds to the centre of the 5-dimensional hypercube and the middle point of the input diagonal. The distribution of $g(\mathbf{x}_0)$ behaves similarly for the Friedman data set with heteroskedastic noise (not shown), and other experiments were conducted with noise from Student laws showing the same behaviour. It suggests that for ELM ensembles, $g(\mathbf{x})$ and $\hat{f}(\mathbf{x})$ may asymptotically follow a Gaussian distribution. However, dependencies exist between the components of $\boldsymbol{\xi}_m$, and also between the members of the ELM ensemble. Therefore, the classical central limit theorem is not directly applicable, and it does not seem straightforward to conclude for Gaussianity in the case of large sample size n or large M. Despite knowing if one can prove or disprove this conclusion, let us assume that the distribution of $g(\mathbf{x})$ is (asymptotically) Gaussian in the remainder of this section. As a matter of fact, $g(\mathbf{x})$ has a unit variance, but it is not centred due to the bias of $\hat{f}(\mathbf{x})$. Its mean is given by

$$\mathbb{E}\left[g(\mathbf{x})\right] = \frac{\mathrm{Bias}[\hat{f}(\mathbf{x})]}{\mathrm{sd}[\hat{f}(\mathbf{x})]},$$

and is reported in Fig. 6.3 as a vertical dashed black line. Obviously, this quantity is unknown in practice but necessary to build a reliable CI for $f(\mathbf{x})$. However, if the bias of $\hat{f}(\mathbf{x})$ is negligible relatively to its variance, then $g(\mathbf{x})$ is close to being centred, and approximate point-wise CI can be derived based solely on an estimation of the variance of $\hat{f}(\mathbf{x})$. That is, if $g(\mathbf{x})$ is close to centred, then the estimated ± 1.96 standard-error around $\hat{f}(\mathbf{x})$ define an approximate point-wise 95% CI for $f(\mathbf{x})$.

Figure 6.4 plots for some of the previous experiments the coverage probability of the approximate point-wise 95% CI, i.e. the proportion of time that the estimated confidence interval actually contains the true $f(\mathbf{x})$ among all the 1′000 experiment repetitions. Figure 6.4 (left) shows the one-dimensional case for $M = 10$. The black dashed line indicates the proportion of time that $f(x)$ lies within $\hat{f}(x) \pm 1.96\,\mathrm{sd}[\hat{f}(x)]$, computed on the basis of the 10′000 simulations baseline. Observe that around each point where the bias vanishes—seen in Fig. 6.1 (left)—this proportion comes closer to the

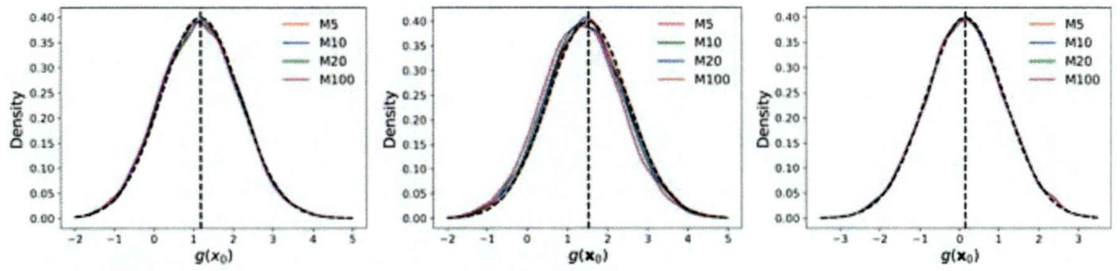

Fig. 6.3 **KDE of** $g(\mathbf{x}_0)$**:** (left) For the one-dimensional case, at $x_0 = \pi/4$; (middle) For the multi-dimensional case with homoskedastic noise, at $\mathbf{x}_0 = 1/2 \cdot (1, 1, 1, 1, 1)$; (right) At the same point with regularised ELM ensemble of $N = 300$ neurons. In black dashed line, Gaussian distributions with unit variance and mean (vertical dashed line), determined by the ratio between the bias and the standard deviation of $\hat{f}(\mathbf{x}_0)$

true coverage probability of 0.95 defined by the confidence level. Conversely, for instance at $x_0 = \pi/4$—which is near the point with the smallest variance, see Fig. 6.1 (right)—the bias is high relative to the variance, then $g(\mathbf{x}_0)$ is far from being centred, which implies a wrong construction of the CI, leading to a bad result. Obviously, $\mathrm{sd}[\hat{f}(x)]$ is not available in practice, and looking at the actual coverage probability of the estimated CI $\hat{f}(x) \pm 1.96\,\hat{\sigma}_{\mathrm{BR}}(x)$ is more interesting. However, note that it reproduces quite fairly the same behaviour, as expected.

The non-regularised multi-dimensional experiment for $M = 5$ is also displayed in Fig. 6.4 for homoskedastic case, also estimated with $\hat{f}(x) \pm 1.96\,\hat{\sigma}_{\mathrm{BR}}(x)$ (middle), and for heteroskedastic case, estimated with $\hat{f}(x) \pm 1.96\,\hat{\sigma}_{S2}(x)$ (right). For both, the actual coverage probability is globally greater than the proportion based on the theoretical CI build with the true $\mathrm{sd}[\hat{f}(x)]$. This is partially explained by the overestimation of the $\hat{f}(\mathbf{x})$ standard deviation—see Sect. 6.5.2. In some low bias regions, this results in slightly conservative CI, i.e. the actual coverage probability is greater than the true coverage probability of 95%. Observe that around $\mathbf{x}_0 = 1/2 \cdot (1, 1, 1, 1, 1)$ the actual coverage probability is especially bad for the homoskedastic case, while it is quite reasonable for the heteroskedastic case. It is explained by the fact that the heteroskedastic noise variance around the centre of the hypercube is up to five times more than the homoskedastic variance, which implies an increase of the variance of $\hat{f}(\mathbf{x})$ and a decrease of $\mathbb{E}[g(\mathbf{x})]$ around $\mathbf{x}_0$. Finally, the actual coverage probability is quite satisfying for the heteroskedastic case.

The effectiveness of the CI estimation for $f(\mathbf{x})$ is highly dependant on the data set at hand, and significant bias of $\hat{f}(\mathbf{x})$ relatively to its variance can lead to highly permissive and bad CI for $f(\mathbf{x})$. However, it is possible to identify potential paths to overcome this problem. Firstly, note that even if the bias of $\hat{f}(\mathbf{x})$ is too important to be ignored, the estimated ± 1.96 standard-error bands around $\hat{f}(\mathbf{x})$ still provides a reliable CI for $\mathbb{E}[\hat{f}(\mathbf{x})]$. Secondly, the bias could be estimated. Thirdly, a manner of reducing the bias is to smooth $\hat{f}(\mathbf{x})$ slightly less than what would be appropriate [31], for instance, through regularisation. For the latter, one provides here an example for the multi-dimensional case with homoskedastic noise.

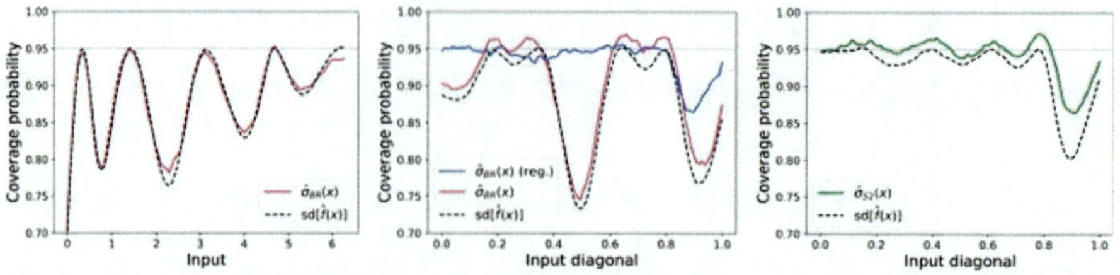

Fig. 6.4 Coverage probabilities: (left) For the one-dimensional case with $M = 10$; (middle) For the multi-dimensional case with $M = 5$, with homoskedastic noise; (right) For the multi-dimensional case with $M = 5$, with heteroskedastic noise. The true coverage probability is fixed at 95% (dotted line). The actual coverage probability is reported for $\hat{f}(x) \pm 1.96\,\hat{\sigma}_{\mathrm{BR}}$ (in purple) and $\hat{f}(x) \pm 1.96\,\hat{\sigma}_{S2}$ (in green). It is also shown for $\hat{f}(x) \pm 1.96\,\mathrm{sd}[\hat{f}(x)]$ in black dashed line, for comparison purpose. In the homoskedastic multi-dimensional case, the actual coverage probability is also reported for $\hat{f}(x) \pm 1.96\,\hat{\sigma}_{\mathrm{BR}}$ for an ensemble of $M = 5$ regularised ELM of $N = 300$ neurons (in blue)

Ensemble of $M = 5$ regularised ELM is trained with $N = 300$. Selecting intentionally a bigger number of neurons increases the model complexity, then reduces the model bias. Nevertheless, increasing complexity also implies increasing the model variability, which puts the model in an overfitting situation that the regularisation mechanism controls at the expense of introducing an additional bias. The GCV estimate gives a Tikhonov factor of 10^{-4}, which introduces too much bias. Then, $\alpha = 10^{-6}$ is empirically set to decrease the amount of smoothing hence alleviating the bias at the expense of higher variance. In order to measure the predictive performance of the model, the Relative mean squared Error (RE) is defined on the training set by

$$\mathrm{RE} = \frac{\frac{1}{n}\sum_{i=1}^{n}\left(y_i - \hat{f}(\mathbf{x}_i)\right)^2}{\frac{1}{n}\sum_{i=1}^{n}\left(y_i - \frac{1}{n}\sum_{i=j}^{n} y_j\right)^2},$$

and a similar measure is defined for the testing set. Generally speaking, lower values of RE are better. A value higher than 1 for RE indicates that the model performs worse than the mean [32]. Note also that RE can be interpreted as an estimation of the ratio between the residual variance and the data variance.

Table 6.4 shows the quantitative results of the regularised model with $N = 300$ compared to the non-regularised model done previously with $N = 91$. On average among the $1'000$ experiments, the testing MSE is slightly better for $N = 91$. However, the testing RE shows it represents 2.8% of the data variance in both cases, and no significant difference is identifiable. As expected, the se_k of the true variance—and its estimation—is greater for $N = 300$. The variance is better estimated, as shown by e_k and re_k. It is likely due to the model bias reduction, which probably implies a decrease of the bias of the noise estimation, and therefore of the bias of the variance estimates—see Sect. 6.4.2. Figure 6.3 (right) visualises the distribution of $g(\mathbf{x}_0)$ at the centre of the 5-dimensional hypercube. Comparing with the first model $N = 91$—

Table 6.4 Non regularised ELM ensemble versus regularised ELM ensemble with increased model complexity. Comparison between a non-regularised model with $N = 91$ neurons and a regularised model with $N = 300$ and $\alpha = 10^{-6}$, on the multi-dimensional synthetic experiment with homoskedastic noise. For both model, $M = 5$ and the variance is estimated with $\hat{\sigma}^2_{\mathrm{BR}}$. Mean (standard deviation) of MSE, RE, se_k, e_k and re_k

		$N = 91$	Grnd tr.	$N = 300$ (reg.)	Grnd tr.
	MSE	0.441 (0.030)	—	0.361 (0.025)	—
	RE	0.017 (0.001)	—	0.014 (0.001)	—
Training set	se_k	0.273 (0.009)	0.255	0.297 (0.010)	0.294
	e_k	0.018 (0.008)	—	0.009 (0.006)	—
	re_k	0.070 (0.032)	—	0.031 (0.022)	—
	MSE	0.679 (0.037)	—	0.682 (0.041)	—
	RE	0.028 (0.002)	—	0.028 (0.002)	—
Testing set	se_k	0.282 (0.010)	0.264	0.312 (0.011)	0.307
	e_k	0.018 (0.008)	—	0.009 (0.007)	—
	re_k	0.069 (0.030)	—	0.031 (0.022)	—

Fig. 6.3 (middle)—the distribution of $g(\mathbf{x}_0)$ is far closer to a centred Gaussian. The coverage probability is also shown and compared in Fig. 6.4 (middle), where results are considerably improved, especially at $\mathbf{x}_0 = 1/2 \cdot (1, 1, 1, 1, 1)$. To summarise, while the CI is globally correctly estimated, the predictive performance is almost the same.

6.6 Summary

Looking back at this chapter, the contribution is threefold. First, analytical developments are proposed to derive ELM variance considering the data noise and the contribution induced by the random input weights and biases. It is done without any other assumptions on the noise distribution than having a vanishing expectation and a finite variance. In particular, the presented theoretical results hold for dependant and non-identically distributed data. The variance of $\hat{f}(\mathbf{x}_0)$ knowing input data has been decomposed into additive terms, supporting the identification and the interpretation of the contribution of different variability sources.

Second, homoskedastic and heteroskedastic variance estimates are provided, and some of their properties are investigated. Note that these estimates are free of the noise distribution. While it may be argued that the homoskedastic case could be unrealistic, its study was of great interest as it provided an insightful propaedeutic value and developed the intuition for more advanced situations. Moreover, in the case of applications with a small number of data, the homoskedastic assumption may yield better results. According to extensive simulations, bias-reduced estimate $\hat{\sigma}^2_{\mathrm{BR}}$ is likely uniformly better than $\hat{\sigma}^2_{\mathrm{NHo}}$ in the homoskedastic case and should be preferred. In the heteroskedastic case, $\hat{\sigma}^2_{S2}$ and $\hat{\sigma}^2_{S3}$ has been empirically shown to be better than other proposed estimates. Although these estimates are close to each other, $\hat{\sigma}^2_{S2}$ is computationally more efficient than $\hat{\sigma}^2_{S3}$.

Third, empirical bases were proposed to move towards CI estimations including the variability induced by the random input weights and biases. Their discussion also raised awareness about some pitfalls of confidence interval estimations.

Note that prediction variance estimation is easily obtained by adding $\hat{\sigma}^2_{\varepsilon}$ to the variance estimate in the homoskedastic case, while the noise variance could be estimated in the heteroskedastic case, e.g. with a second model [8, 9]. Prediction interval can also be constructed, assuming convenient noise distribution.

Overall, the presented results clarified the impact of input weights variability and noise, hence increasing the understanding of ELM variability. These UQ investigations on ELM are developed in the broad context of the non-linear regression problem, and therefore the contribution is quite general. This knowledge will be combined in the next chapter with the spatio-temporal interpolation framework of Chap. 5 to predict wind speed value and estimate its variance at any time and any location.

References

1. Heskes T (1997) Practical confidence and prediction intervals. In: Mozer MC, Jordan MI, Petsche T (eds), Advances in neural information processing systems, vol 9. MIT Press, pp. 176–182. http://papers.nips.cc/paper/1306-practical-confidence-and-predictionintervals.pdf
2. Tibshirani R (1996) A comparison of some error estimates for neural network models. In: Neural computation, vol 8, no 1, pp 152–163. https://doi.org/10.1162/neco.1996.8.1.152. eprint: https://doi.org/10.1162/neco.1996.8.1.152. [Online]. https://doi.org/10.1162/neco.1996.8.1.152
3. Dybowski R, Roberts SJ (2001) Confidence intervals and prediction intervals for feed-forward neural networks. Clinical applications of artificial neural networks, pp 298–326
4. Huang G-B, Zhu Q-Y, Siew C-K (2004) Extreme learning machine: a new learning scheme of feedforward neural networks. In: 2004 IEEE international joint conference on neural networks (IEEE Cat No 04CH37541), vol 2, pp. 985–990, July 2004. https://doi.org/10.1109/IJCNN.2004.1380068
5. Chatfield C (1995) Model uncertainty, data mining and statistical inference. J R Stat Soc: Ser A (Statistics in Society) 158(3):419–444
6. Ning K, Liu M, Dong M (2015) A new robust ELM method based on a bayesian framework with heavy-tailed distribution and weighted likelihood function. Neurocomputing 149:891–903
7. Soria-Olivas E, Gomez-Sanchis J, Martin JD et al (2011) BELM: Bayesian extreme learning machine. IEEE Trans Neural Netw 22(3), 505–509. https://doi.org/10.1109/TNN.2010.2103956
8. Wan C, Xu Z, Pinson P, Dong ZY, Wong KP (2014) Probabilistic forecasting of wind power generation using extreme learning machine. IEEE Trans Power Syst 29(3):1033–1044. https://doi.org/10.1109/TPWRS.2013.2287871 Mar
9. Akusok A, Miche Y, Björk K-M, Lendasse A (2019) Per-sample prediction intervals for extreme learning machines. Int J Mach Learn Cybern 10(5):991–1001. ISSN: 1868- 808X. https://doi.org/10.1007/s13042-017-0777-2.[Online]. https://doi.org/10.1007/s13042-017-0777-2
10. Leuenberger M, Kanevski M (2015) Extreme learning machines for spatial environmental data. Comput Geosci 85:64–73
11. Huang G, Huang G-B, Song S, You K (2015) Trends in extreme learning machines: a review. Neural Netw 61:32–48
12. Huang G-B, Zhu Q-Y, Siew C-K (2006) Extreme learning machine: theory and applications. Neurocomputing 70(1–3):489–501
13. Davidson R, MacKinnon JG et al (2004) Econometric theory and methods, vol 5. Oxford University Press, New York
14. Davison AC (2003) Statistical models, ser. Cambridge series in statistical and probabilistic mathematics. Cambridge University Press. https://doi.org/10.1017/CBO9780511815850
15. Deng W, Zheng Q, Chen L (2009) Regularized extreme learning machine. In: IEEE symposium on computational intelligence and data mining. IEEE, pp 389–395
16. Lendasse A, Akusok A, Simula O et al (2013) Extreme learning machine: a robust modeling technique? yes! In: International work-conference on artificial neural networks. Springer, pp 17–35
17. Piegorsch WW (2015) Statistical data analytics: foundations for data mining, in-formatics, and knowledge discovery. Wiley
18. Boyd S, Vandenberghe L (2018) Introduction to applied linear algebra: vectors, matrices, and least squares. Cambridge University Press
19. Liu N, Wang H (2010) Ensemble based extreme learning machine. IEEE Signal Process Lett 17(8):754–757
20. Mathai AM, Provost SB (1992) Quadratic forms in random variables: theory and applications. Dekker
21. Seber GA (2009) Multivariate observations, vol 252. Wiley
22. Hastie TJ, Tibshirani RJ (1990) Generalized additive models, vol 43. CRC Press

23. Golub GH, Heath M, Wahba G (1979) Generalized cross-validation as a method for choosing a good ridge parameter. Technometrics 21(2):215–223
24. MacKinnon JG, White H (1985) Some heteroskedasticity-consistent covariance matrix estimators with improved finite sample properties. J Econom 29(3):305–325
25. Nyquist H (1988) Applications of the jackknife procedure in ridge regression. Comput Stat Data Anal 6(2):177–183
26. White H (1980) A heteroskedasticity-consistent covariance matrix estimator and a direct test for heteroskedasticity. Econometrica: J Econom Soc 817–838
27. Hinkley DV (1977) Jackknifing in unbalanced situations. Technometrics 19(3):285–292
28. Horn SD, Horn RA, Duncan DB (1975) Estimating heteroscedastic variances in linear models. J Am Stat Assoc 70(350):380–385
29. Cribari-Neto F (2004) Asymptotic inference under heteroskedasticity of unknown form. Comput Stat Data Anal 45(2):215–233
30. Friedman JH (1991) Multivariate adaptive regression splines. Ann Statist 19(1):1–67. https://doi.org/10.1214/aos/1176347963
31. Hall P (1992) Effect of bias estimation on coverage accuracy of bootstrap confidence intervals for a probability density. Ann Stat 675–694
32. Golay J, Leuenberger M, Kanevski M (2017) Feature selection for regression problems based on the morisita estimator of intrinsic dimension. Pattern Recog 70:126–138

Chapter 7
Spatio-Temporal Modelling Using Extreme Learning Machine

This chapter combines the developments of Chaps. 5 and 6 by using ELM to model each spatial coefficient map obtained from the EOF decomposition and extend the UQ to the spatio-temporal prediction. The modelling variance of the spatio-temporal prediction is estimated by reusing all component ELM variances obtained by the method presented in the previous chapter. The prediction variance is estimated by interpolating the expected squared residuals with a second model. However, this second modelling is performed on log-transformed data to ensure positive variance estimation, which generates the need to know the transformed variable's variance. This issue is solved thanks again to ELM variance estimation. An application on the MeteoSwiss wind speed data is presented, providing an estimation of the power potential of aeolian energy in rural areas of Switzerland. The spatio-temporal ELM model, its model variance and prediction variance estimations are detailed in Sect. 7.1. The model is then applied to the MeteoSwiss data to estimate the wind speed field in Switzerland for ten years. It is presented in Sect. 7.2 where detailed residuals analysis is also performed. Section 7.3 proposes to convert the estimated wind speed field into renewable power potential and propagate its uncertainty.

7.1 Spatio-Temporal ELM Model

This section first discusses the combination of the proposed spatio-temporal framework with ELM. Subsequently, the variance of the spatio-temporal model is derived and estimated. Finally, the prediction variance is estimated by using a second spatio-temporal model.

F. Guignard, *On Spatio-Temporal Data Modelling and Uncertainty Quantification Using Machine Learning and Information Theory* , Springer Theses,
https://doi.org/10.1007/978-3-030-95231-0_7

7.1.1 ELM Modelling of the Spatial Coefficients

Recall the spatio-temporal model given in Eq. (5.1). Each coefficient $a_k(\mathbf{s}_i) = a_k(\mathbf{s}_i, \mathbf{x}_i)$ depends only on space and potentially through additional spatial features $\mathbf{x}(\mathbf{s}_i)$. The additional spatial features are dropped to lighten the notation, which is still consistent as they also depend on $\mathbf{s}_i$. Using the single output strategy presented in Chap. 5, the coefficient maps are modelled with ELM, although they could be modelled with any ML algorithm. For the kth stochastic coefficient map, this implicitly assumes that a deterministic function f_k is sought such that

$$a_k(\mathbf{s}_i) = f_k(\mathbf{s}_i) + \varepsilon_k(\mathbf{s}_i), \tag{7.1}$$

where $\varepsilon_k(\mathbf{s}_i)$ is a centred random noise with finite variance. The function estimation is noted $\hat{f}_k(\mathbf{s}_i) = \hat{a}_k(\mathbf{s}_i)$ and is used as a predicted spatial coefficient map. The spatio-temporal prediction at a new interpolated point is given by

$$\widehat{Z}(\mathbf{s}_0, t_j) = \widehat{\mu}_t(t_j) + \sum_{k=1}^{K} \hat{a}_k(\mathbf{s}_0)\phi_k(t_j). \tag{7.2}$$

By subtracting Eq. (5.1) by Eq. (7.2), the prediction error is then

$$\begin{aligned} Z(\mathbf{s}_0, t_j) - \widehat{Z}(\mathbf{s}_0, t_j) &= \sum_{k=1}^{K} \left[a_k(\mathbf{s}_0) - \hat{a}_k(\mathbf{s}_0)\right] \phi_k(t_j) + \eta(\mathbf{s}_0, t_j) \\ &= \underbrace{\sum_{k=1}^{K} \left[f_k(\mathbf{s}_0) - \hat{f}_k(\mathbf{s}_0)\right] \phi_k(t_j)}_{\text{modelling error}} + \sum_{k=1}^{K} \varepsilon_k(\mathbf{s}_0)\phi_k(t_j) + \eta(\mathbf{s}_0, t_j). \end{aligned}$$

The first term in the latter equation is the modelling error between the spatio-temporal linear combination of spatial estimates $\hat{f}_k(\mathbf{s}_0)$ and the linear combination of the true regression functions $f_k(\mathbf{s}_0)$.

7.1.2 Model Variance Estimation

The variance of the modelling error, noted $\sigma_C^2(\mathbf{s}_0, t_j)$, will be referred to as *model variance*. It quantifies the model accuracy. Let $\mathbf{y}_k$ denote the vector of training output where $a_k(\mathbf{s}_i)$ is the ith vector component. Analogously, $\boldsymbol{\varepsilon}_k$ denote the vector given by the noise at the training points $\varepsilon_k(\mathbf{s}_i)$. Assume that $\text{Cov}\left[\boldsymbol{\varepsilon}_k,\ \boldsymbol{\varepsilon}_l\right] = 0$, or more generally that the component noise is pairwise independent. It allows us to ensure that no additional variability comes from the spatial model interactions. Indeed, knowing the training input, note that for a single ELM and all $k \neq l$,

$$\begin{aligned}
\mathrm{Cov}\left[\hat{f}_k(\mathbf{s}_0),\ \hat{f}_l(\mathbf{s}_0)\right] &= \mathrm{Cov}\left[\mathbf{h}_k^T\mathbf{H}_k^\alpha\mathbf{y}_k,\ \mathbf{h}_l^T\mathbf{H}_l^\alpha\mathbf{y}_l\right] \\
&= \mathrm{Cov}\left[\mathbf{h}_k^T\mathbf{H}_k^\alpha\mathbb{E}\left[\mathbf{y}_k\right],\ \mathbf{h}_l^T\mathbf{H}_l^\alpha\mathbb{E}\left[\mathbf{y}_l\right]\right] \\
&\quad + \mathbb{E}\left[\mathbf{h}_k^T\mathbf{H}_k^\alpha\mathrm{Cov}\left[\mathbf{y}_k,\ \mathbf{y}_l\right]\mathbf{H}_l^{\alpha T}\mathbf{h}_l\right] \\
&= \mathbb{E}\left[\mathbf{y}_k\right]^T\mathrm{Cov}\left[\mathbf{h}_k^T\mathbf{H}_k^\alpha,\ \mathbf{h}_l^T\mathbf{H}_l^\alpha\right]\mathbb{E}\left[\mathbf{y}_l\right] \\
&\quad + \mathbb{E}\left[\mathbf{h}_k^T\mathbf{H}_k^\alpha\mathrm{Cov}\left[\boldsymbol{\varepsilon}_k,\ \boldsymbol{\varepsilon}_l\right]\mathbf{H}_l^{\alpha T}\mathbf{h}_l\right] \\
&= 0,
\end{aligned}$$

where the law of total covariance is used in the second equality. From the latter equations, it is straightforward to check that the covariance between components at $\mathbf{s}_0$ also vanishes for the ELM ensemble. Hence, for ELM ensemble, as the function basis is considered as fixed, one has

$$\mathrm{Cov}\left[\left[f_k(\mathbf{s}_0) - \hat{f}_k(\mathbf{s}_0)\right]\phi_k(t_j),\ \left[f_l(\mathbf{s}_0) - \hat{f}_l(\mathbf{s}_0)\right]\phi_l(t_j)\right] = 0. \tag{7.3}$$

Although it may seems reasonable to assume $\mathrm{Cov}\left[\boldsymbol{\varepsilon}_k,\ \boldsymbol{\varepsilon}_l\right] = 0$, this should be checked, e.g. by looking at the empirical cross-covariance function or cross-variogram of the training residuals.

The model variance can now be computed. Using Eq. (7.3), one obtains

$$\begin{aligned}
\sigma_C^2(\mathbf{s}_0, t_j) &= \mathrm{Var}\left[\sum_{k=1}^K\left[f_k(\mathbf{s}_0) - \hat{f}_k(\mathbf{s}_0)\right]\phi_k(t_j)\right] \\
&= \sum_{k=1}^K \mathrm{Var}\left[f_k(\mathbf{s}_0)\phi_k(t_j) - \hat{f}_k(\mathbf{s}_0)\phi_k(t_j)\right] \\
&= \sum_{k=1}^K \mathrm{Var}\left[\hat{f}_k(\mathbf{s}_0)\right]\phi_k^2(t_j).
\end{aligned}$$

Thus, the spatio-temporal model variance can be obtained directly by a sum of the spatial component model variances weighted by the corresponding squared basis function. Moreover, if we accept the asymptotic Gaussianity of $\hat{f}_k(\mathbf{s}_0)$—which seems quite reasonable, see Sect. 6.5.3—then CI for the spatio-temporal interpolation model could be considered, as long as proper management of the bias is carried on. Hence, the model variance $\sigma_C^2(\mathbf{s}_0, t_j)$ is naturally estimated by using the variance estimate of each ELM ensemble model, by

$$\hat{\sigma}_C^2(\mathbf{s}_0, t_j) = \sum_{k=1}^K \hat{\sigma}_{S2,k}^2(\mathbf{s}_0)\phi_k^2(t_j),$$

where $\hat{\sigma}_{S2,k}^2(\mathbf{s}_0)$ is the heteroskedastic variance estimate $\hat{\sigma}_{S2}^2$ of the ELM-based modelled regression function of spatial coefficient map of the kth EOF component, at the input point $(\mathbf{s}_0, \mathbf{x}(\mathbf{s}_0))$. This choice is motivated by the nice trade-off between estimation effectiveness and computational efficiency of $\hat{\sigma}^2_{S2}$; see Sects. 6.5 and 6.6 of Chap. 6.

7.1.3 Prediction Variance Estimation

The *prediction variance*, given by the variance of the prediction error,

$$\sigma_P^2(\mathbf{s}_0, t_j) = \mathrm{Var}\left[Z(\mathbf{s}_0, t_j) - \widehat{Z}(\mathbf{s}_0, t_j)\right],$$

evaluate the accuracy of the estimate regarding the observed output. Prediction variance encloses the corresponding model variance [1]. A common approach to obtain the conditional variance function $\sigma_P^2(\mathbf{s}_0, t_j)$ is to model it as a function of the input features using the squared residuals [2]. It is motivated by the fact that the expectation of squared residuals approximates the prediction variance [3] and performing a regression on squared residuals gives a reasonable estimate of it [4].

The (stochastic) training squared residuals are given by

$$R^2(\mathbf{s}_i, t_j) = \left(Z(\mathbf{s}_i, t_j) - \widehat{Z}(\mathbf{s}_i, t_j)\right)^2,$$

that will be also noted R^2 for short. The training squared residuals can be used as a training set to a new model. However, this new model may provide a negative estimate of $\sigma_P^2(\mathbf{s}_0, t_j)$. The positiveness of the modelled conditional variance function is ensured through exponentiation, as often proposed [1, 2]. The logarithm values of the squared training residuals of the first model are used as a new training set to model the random variable $L = L(\mathbf{s}_0, t_j) = \log(R^2(\mathbf{s}_0, t_j))$ with mean $\mu_L(\mathbf{s}_0, t_j)$ and variance $\sigma_L^2(\mathbf{s}_0, t_j)$, all depending on space and time. This second spatio-temporal model follows the same procedure as before, through the EOF decomposition process and ELM modelling on each component with the high-dimensional input space composed by the spatial features. Its predicted value is noted $\widehat{L}(\mathbf{s}_0, t_j)$.

Getting back the expected squared residuals from their log-transform would produce a bias due to the non-linearity of the exponential function. To partially correct this bias, the following second-order Taylor expansion around μ_L,

$$\exp(L) \simeq \exp(\mu_L) + \exp(\mu_L)(L - \mu_L) + \frac{1}{2}\exp(\mu_L)(L - \mu_L)^2.$$

The expansion of a function of a random variable around its expectation is known in statistics as the *delta method* [5, 6]. Taking the expectation on both sides, one finds

$$
\begin{aligned}
\mathbb{E}\left[R^2\right] &= \mathbb{E}\left[\exp(L)\right] \\
&\simeq \exp(\mu_L) + \frac{1}{2}\exp(\mu_L)\,\mathbb{E}\left[(L-\mu_L)^2\right] \\
&= \exp(\mu_L)\left(1+\frac{1}{2}\sigma_L^2\right).
\end{aligned}
$$

This motivates the following estimation of the prediction variance,

$$
\hat{\sigma}_P^2(\mathbf{s}_0, t_j) = \exp\left(\hat{\mu}_L\right)\left(1+\frac{1}{2}\hat{\sigma}_L^2\right),
$$

with the prediction of the second spatio-temporal model $\hat{\mu}_L = \widehat{L}(\mathbf{s}_0, t_j)$ and its prediction variance estimate

$$
\begin{aligned}
\hat{\sigma}_L^2 = \hat{\sigma}_L^2(\mathbf{s}_0, t_j) &= \sum_{k=1}^{K}\left[\hat{\sigma}_{BR,k}^2(\mathbf{s}_0) + \hat{\sigma}_{\varepsilon,k}^2\right]\phi_k^2(t_j) \\
&= \sum_{k=1}^{K}\hat{\sigma}_{BR,k}^2(\mathbf{s}_0)\phi_k^2(t_j) + \sum_{k=1}^{K}\hat{\sigma}_{\varepsilon,k}^2\phi_k^2(t_j),
\end{aligned}
$$

where $\hat{\sigma}_{BR,k}^2(\mathbf{s}_0)$—respectively $\hat{\sigma}_{\varepsilon,k}^2$—is the bias-reduced homoskedastic variance estimate $\hat{\sigma}_{BR}^2$—respectively the noise estimate $\hat{\sigma}_\varepsilon^2$—of the kth modelled spatial coefficient map of the second spatio-temporal model. Although the noise of each component is not necessary homoskedastic, $\hat{\sigma}_L^2(\mathbf{s}_0, t_j)$ furnish a good guess to estimate $\sigma_L^2(\mathbf{s}_0, t_j)$ and is better than limited the estimation of $\sigma_P^2(\mathbf{s}_0, t_j)$ to a first-order Taylor expansion.

7.2 Application to the MeteoSwiss Data

The methodology proposed in the previous section is applied to the Meteoswiss data presented in Chap. 2. The residuals are then analysed, and data are compared before and after modelling with various EDA tools.

7.2.1 Wind Speed Modelling

The framework is applied to the MSWind 08–12, MSWind 13–16 and MSWind 17 data sets. For both the first and the second spatio-temporal models, the coefficients of each EOF component are spatially modelled with a regularised ELM ensemble of $M = 20$ members by using the 13-dimensional input space presented in Sect. 2.1. The number of neurons of each ELM is fixed within each ensemble but changes across the data sets to be slightly smaller than the number of training stations, providing high flexibility to the model. The number of neurons is given in Table 7.1 for all data set. Then, each member of each ELM ensemble is regularised. The Tikhonov factor α is selected by GCV [7, 8]. Then the spatio-temporal wind speed field, its model variance and prediction variance are modelled on a 250 [m] resolution regular grid.

Considering regularised ELM as a ridge regression [9, 10] in a random feature space provides a powerful interpretation in the context of these spatio-temporal models. Tikhonov factors for the first 25 EOF components of each model are presented in Fig. 7.1 as matrices with a linear colour scale depending on the magnitude order of α. For instance, a factor in yellow indicates the selection of a huge regularisation parameter which shrinks the output weights of ELM near zero and suggests that there is not enough structure in the corresponding spatial coefficient map to model it. In this case, the model variance of the interpolated coefficient also tends towards zero. Conversely, a factor in dark tone indicates a regularisation parameter closer to zero and, therefore, a more moderate amount of regularisation closer to a classical LS procedure. It suggests that the regularised ELM is prone to consider a spatial structure for the corresponding EOF coefficients.

Let us first have a general look at Fig. 7.1. Generally speaking, the matrice of the first model tends to be less sparse with more recent data, possibly due to the increasing number of available training stations used for the EOF decomposition. Reading the matrices vertically, the first components provide the main contribution in terms of data variability but also in terms of spatially structured information. For visualisation purposes, components above the 25th component containing less variability are not shown. These latter almost all have a Tikhonov factor at their maximum and then have a minimal and negligible effect on the final recomposition of spatio-temporal fields. In between is a transitional behaviour. Reading now the matrices horizontally, variability within ensembles can be detected for some components of the first component group, while no variability appears in others. For the transitional behaviour, some ELM members within ensembles contribute to the model while others do not.

Let us now have a closer look at the first models. For the MSWind 08-12 data—seen in Fig. 7.1a—it is mainly used the first five components, which from the EOF decomposition are known to cumulate 65% of the variability. Excepted for some isolated members in components 7 and 13, the model uses no other information. For the MSWind 13–16 data—seen in Fig. 7.1b—components 1, 2, 3 and 6 are selected for all members, which are also representing 65% of the variability.It is more sporadic for components 4, 5, 15, 16 and 22. For the MSWind 17 data—seen in Fig. 7.1c—components 1 to 6, 9, 11 and 17 are fully contributing to the model and correspond to 73% of the variability, while components 8, 13, 19, 20 and 24 are only partly contributing. All other components are automatically not considered due to the regularisation mechanism. For the three data sets, the first component of the original data model is variable, denoting some hesitations of the model. It may be because this component contains seasonal cycles which are weakly dependent on space; see Figs. 3.2 and 5.9. For the second model based on log-squared residuals, the behaviour is more contrasted.

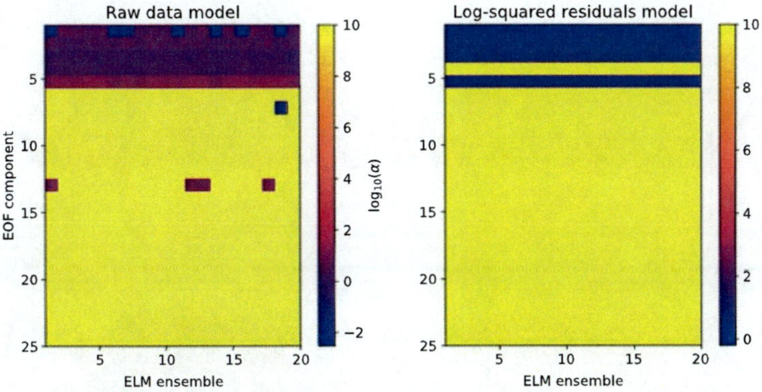

(A) MSWind 08-12.

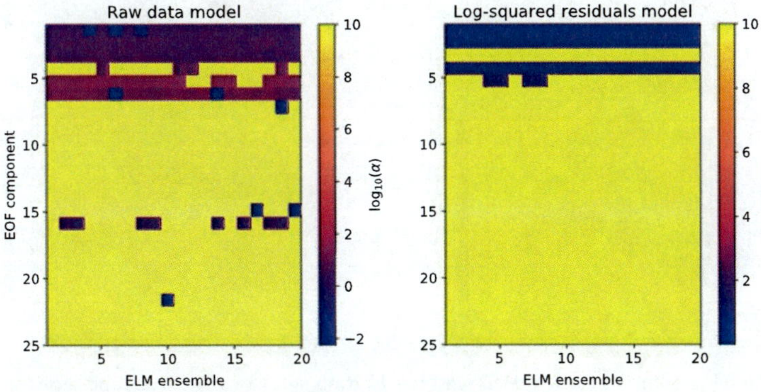

(B) MSWind 13-16.

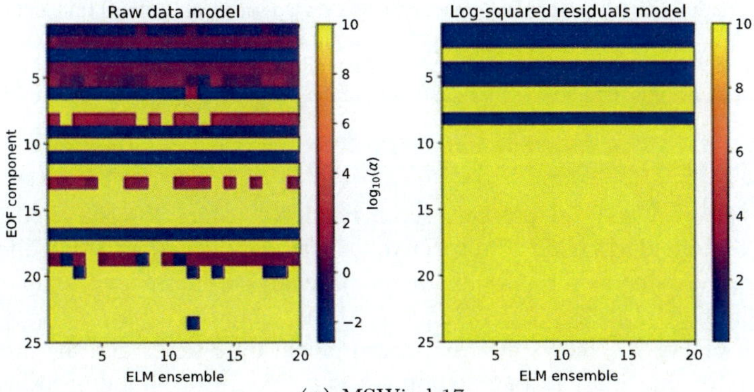

(C) MSWind 17.

Fig. 7.1 Regularisation parameters for the first 25 components, for the 20 members of the ELM ensemble and all models: (left) The first spatio-temporal model is performed on the original wind speed data and provides the interpolated values and its model variance; (right) The second spatio-temporal model is performed on the log-squared residuals of the first one and provides the prediction variance

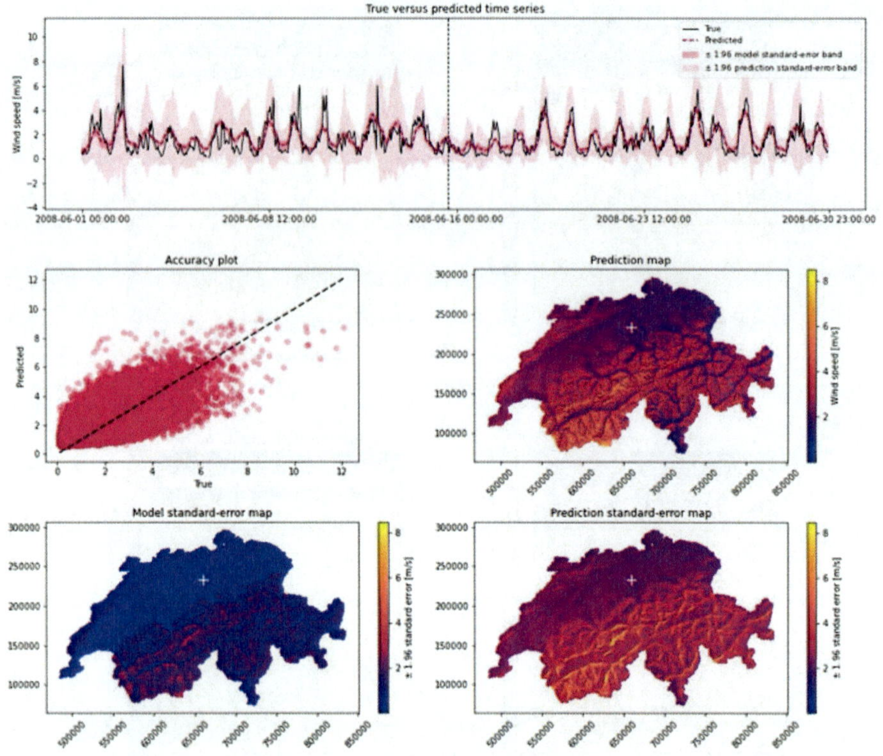

Fig. 7.2 Model prediction for the MSWind 08–12 data set: (top) The true time series (in black) at the MOA testing station marked by a cross in the maps below and the predicted time series (in magenta). For visualisation purposes, only June 2008 is shown; (middle left) Accuracy plot at the same testing station; (middle right) The predicted map of wind speed, at the fixed time indicated by the vertical dashed line in the temporal plot above; (bottom left) Map of the model standard error multiplied by 1.96 at the same time; (bottom right) Map of the prediction standard error multiplied by 1.96 at the same time, obtained as the square root of the output from the second spatio-temporal model

Figures 7.2, 7.3 and 7.4 presents some predicted values for MSWind 08-12, MSWind 13-16 and MSWind 17 respectively. Note that the periods studied in these Figures are the same as the ones explored and analysed in previous chapters and visualised in Figs. 2.11, 3.1, 3.2, 3.3, 5.9, 5.10 and 5.11.

The top of Fig. 7.2 shows the true measurement time series and the predicted time series for the MOA testing station (Mosen, 453 m of altitude) during June 2008. The model reproduces the main characteristics of the measured wind speed time series. It replicates the daily cyclicity detected in Figs. 2.11 and 3.3 as well as most of the changes of magnitude and behaviour. However, a smoothing effect is also observed in the predicted time series. Estimation of pointwise model and prediction standard-error bands based respectively on $\pm 1.96\,\hat{\sigma}_C$ and $\pm 1.96\,\hat{\sigma}_P$ are also reported. The model standard-error band is relatively narrow, suggesting a low

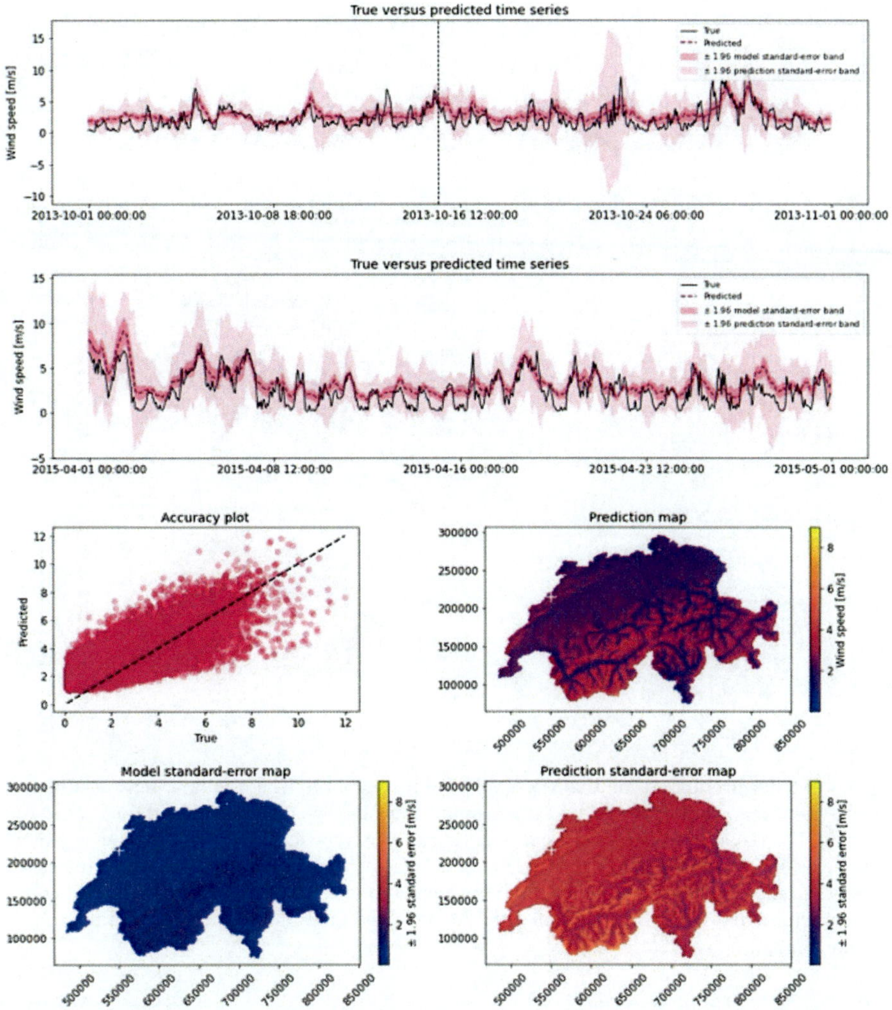

Fig. 7.3 **Model prediction for the MSWind 13–16 data set**: (first and second rows) The true time series (in black) at the CDF testing station marked by a cross in the maps below and the predicted time series (in magenta). For visualisation purposes, only October 2013 for the first plot and April 2015 for the second plot are shown; (third row left) Accuracy plot at the same testing station; (third row right) The predicted map of wind speed, at the fixed time indicated by the vertical dashed line in the temporal plot above; (fourth row left) Map of the model standard error multiplied by 1.96 at the same time; (fourth row right) Map of the prediction standard error multiplied by 1.96 at the same time

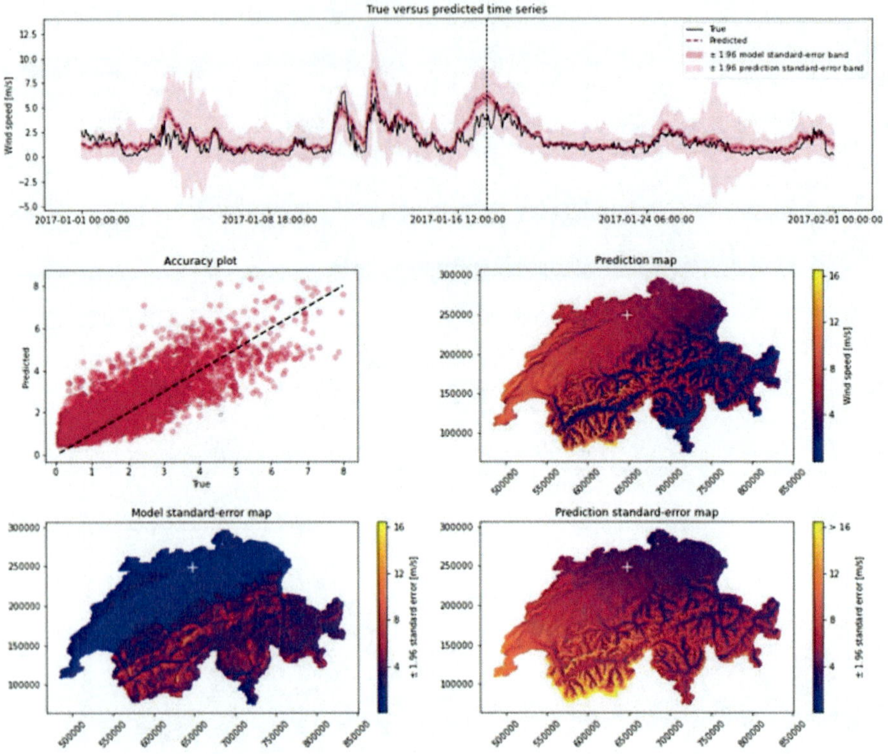

Fig. 7.4 **Model prediction for the MSWind 17 data set**: (top) The true time series (in black) at the BUS testing station marked by a cross in the maps below and the predicted time series (in magenta). For visualisation purposes, only January 2017 is shown; (middle left) Accuracy plot at the same testing station; (middle right) The predicted map of wind speed, at the fixed time indicated by the vertical dashed line in the temporal plot above; (bottom left) Map of 1.96 times the model standard error at the same time; (bottom right) Map of 1.96 times the prediction standard error at the same time

variability of the mean prediction, despite the low number of training stations. On the contrary, the prediction standard-error band is rather large, as expected from the noisy nature of wind speed data. It is interesting to remark that the true wind speed time series is well encompassed in the ± 1.96 prediction standard-error bands. For the fixed time marked in the time series plot, a predicted wind speed map is displayed in the same Figure. It shows a lower wind speed on the Plateau, where the MOA station is located, which is expected. At the same fixed time, model and prediction standard-error maps are visualized. For both maps, a higher uncertainty is seen in the Alps, which could be explained by a higher variability of the data in this region. However, in the model standard-error map, this pattern may also be generated because of the weak representativity of some features in the 13-dimensional input space; see Sect. 2.2.2.

Figures 7.3 and 7.4 globally show analogous results for the two other data sets and general comments done on Fig. 7.2 still applied. The first and second rows of Fig. 7.3 shows the true and predicted time series for the CDF testing station (La Chaux-de-Fonds, 1017 m of altitude) during October 2013 and April 2015, respectively. While the prediction is quite satisfying for October 2013, it seems to overestimate true wind speed slightly for April 2015. However, this also results in larger ± 1.96 prediction standard-error bands which still well encompasses the true wind speed. This also suggests that the uncertainty quantification estimation based on the log-squared residuals does consistent work. The three maps seem to have a less marked pattern. Figure 7.4 displays the BUS station—already shown in Chap. 5 for the DL-based modelling assessment—during January 2017. By comparing with Fig. 5.10, the ELM approach provides smoother time series prediction, probably because of the automatic discarding of EOF components with spatially unstructured coefficient map. The three geographical regions are easily identifiable on the three maps of the spatial snapshot. Again, a higher model and prediction variabilities are observed in the Alps. Qualitatively, it seems that the spatial scale of the pattern seen on the prediction standard-error map is comparable to the one observed on the prediction map, while the spatial scale of the pattern seen on the model standard-error map is coarser. It may be related to the multi-scale features used in the 13-dimensional space. The three Figures also displays an accuracy plot of the corresponding visualised testing station. It seems that their behaviour improves with more recent data sets, which may be related to the increasing number of training stations, although they highly depend on the studied station.

Comparing the time series of Figs. 7.2, 7.3 and 7.4 with the corresponding empirical temporal mean times series of Fig. 2.11 suggests that the predicted time series are rather close to the temporal mean value, although the true time series are also near to it. Some patterns in the prediction variance could also be recognised in the variability of the training stations displayed in Fig. 2.11. Someone might say that one could simply predict wind speed time series by computing the mean across all stations. Actually, it might work well only for some periods, and it is what suggests the weak spatial structures described in Sect. 3.2. However, space matters during periods with a higher spatial correlation like January. Moreover, the model could also find structures via the other features of the 13-dimensional space, even during periods with weak spatial correlation.

To confirm that the spatio-temporal model does better on the wind speed data, its testing RMSE and MAE are computed and reported in Table 7.1 and compared with the empirical temporal mean time series $\widehat{\mu}_t(t_j)$. Let us specify that $\widehat{\mu}_t(t_j)$ is computed from the training data and used as a prediction to the testing stations, from which testing RMSE and MAE are computed and used as a benchmark. These results show that the spatio-temporal model furnishes a substantial prediction improvement. The results can also be compared to the DL-based approach applied to the MSWind 17 data set in Sect. 5.5, which provides slightly better prediction performances than the ELM-based model—RMSE 1.845, MAE 1.240. However, the DL approach provided no uncertainty assessment.

Table 7.1 Number of neurons and testing RMSE and MAE. Metrics are computed on the set of testing stations. The results based on the spatio-temporal ELM (ST ELM) model are benchmarked to the empirical temporal mean (Emp. temp. mean) time series $\widehat{\mu}_t(t_j)$

Data set	N	ST ELM model		Emp. temp. mean	
		RMSE	MAE	RMSE	MAE
MSWind 08-12	80	2.161	1.429	2.760	1.730
MSWind 13-16	100	1.744	1.282	2.060	1.490
MSWind 17	150	1.929	1.328	2.251	1.513

7.2.2 *Residual Analysis*

A careful analysis of the residuals is carried on. Figure 7.5 displays histograms of training and testing set of the raw data, modelled data and residuals, for the three periods of study. Note that the plots are zoomed in for visualisation purposes, and the actual ranges are reported in Table 7.2, also with the empirical means. The models predict negative values for the three data sets. Although a negative wind speed has no physical meaning, histograms show that it is happening quite rarely. These predicted values will be set to zero for power estimation in the next section. The training residual means are all null, and testing ones are close to zero. However, each residual distribution has its mode below zero and is slightly skewed.

Spatio-temporal variography analysis is performed on the training data for the discussed months. Figure 7.6 visualise the semivariograms for raw data, model predictions and their residuals. The model reproduces well the spatio-temporal dependencies detected by variography for the selected months. As already observed with the DL model in Chap. 5, the sill is lower for the modelled data, suggesting a substantial variability loss, likely due to the noisy nature of the raw data. The semivariograms of the residuals are close to flat with a residual temporal structure. Almost a pure nugget effect is observed along the spatial axis, although a residual structure could subsist for January 2017. Globally, the patterns observed in the raw data semivariograms are reproduced in the corresponding semivariograms of modelled data, modulo the sill shift. It is even more striking by looking at longer temporal lags, e.g. for January 2017, where a high similarity is observed between the semivariogram shapes; see Fig. 7.7.

Spatial variography [11] is also performed on spatial coefficients of some EOF components before and after ELM modelling. Figure 7.8 displays such analysis for the MSWind 17 data set for the first three components, which contain most of the variability; see Sect. 5.5. The (spatial) omnidirectional semivariogram is computed on the spatial coefficient obtained directly from the EOF decomposition of the original spatio-temporal data (in solid line). It shows the presence of the spatial structures in the first three components, although it is less pronounced for the very first component, which is consistent with what was observed and discussed so far in Fig. 7.1.

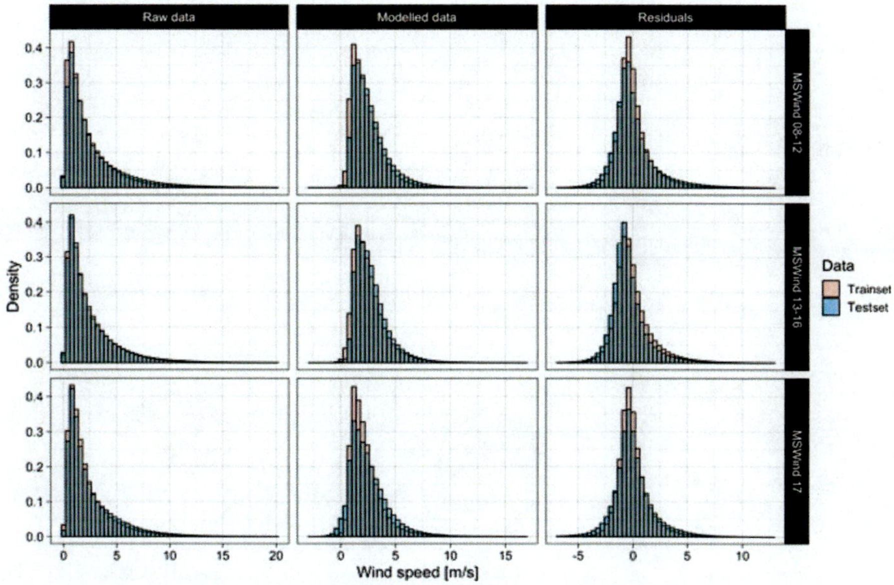

Fig. 7.5 Histograms. For visualisation purposes, the plots are zoomed in. The actual ranges and the means of each histogram are reported in Table 7.2

Table 7.2 Summary statistics. The minimum, the empirical mean, and the maximum are computed on the original data before modelling (Raw), the modelled data (Mod.) and residuals (Res.) for the training and testing stations of the three data sets

		Training set			Testing set		
		Raw	Mod.	Res.	Raw	Mod.	Res.
MSWind 08-12	Min.	0.10	-0.97	-9.08	0.10	-0.25	-11.83
	Mean	2.33	2.33	0.00	2.81	2.71	0.10
	Max.	38.08	20.22	22.13	42.85	18.30	34.54
MSWind 13-16	Min.	0.10	-1.26	-11.62	0.10	-0.71	-10.11
	Mean	2.56	2.56	0.00	2.42	2.78	-0.37
	Max.	40.43	19.59	29.22	25.13	16.88	16.89
MSWind 17	Min.	0.10	-2.88	-8.66	0.10	-2.69	-17.18
	Mean	2.30	2.30	0.00	2.56	2.50	0.06
	Max.	34.15	26.32	24.44	34.67	20.35	25.04

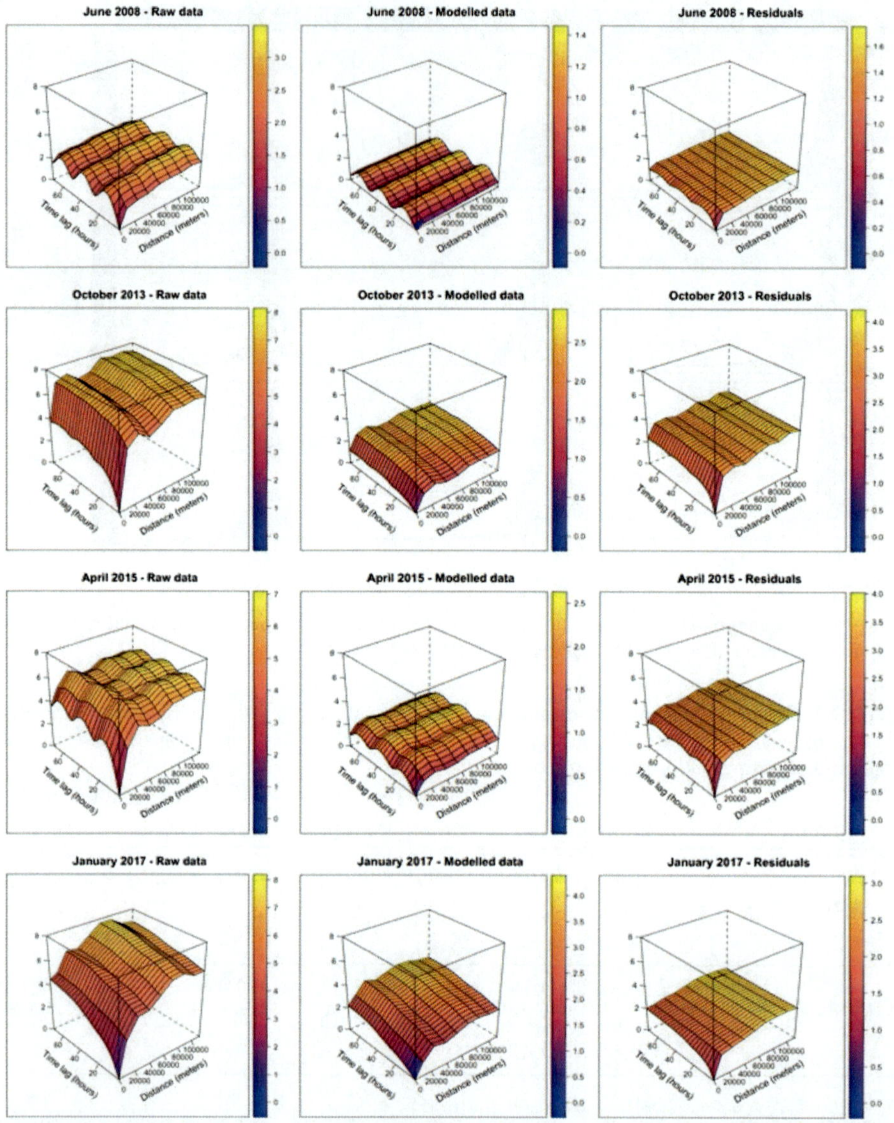

Fig. 7.6 Variography for the spatio-temporal ELM modelling. Spatio-temporal semivariograms have been computed on the training points for different months. The corresponding sample variances can be found in Table 7.3

Figure 7.8 also shows the semivariograms of the residuals obtained from the spatial modelling with ELM ensembles (in dashed line), which are close to pure nugget effects. It has two consequences. First, the spatial modelling component by component seems to extract the spatial structures correctly. Second, this suggests that the heteroskedastic variance estimate of the ELM ensembles satisfies its independence assumption—actually, it is sufficient to assume a vanishing covariance for this estimation—and hence is appropriate. Finally, the cross-variograms between the residuals of component pairs fluctuate near zero (in dashed-dotted line). Compared to the corresponding semivariograms, this indicates that the correlation between component residual pairs is very weak [12] and then satisfy the additional assumption for model variance estimation, which is necessary to ensure that no variability comes from model interactions. The other components, containing far less variability, can be exempted from such analysis without too many risks.

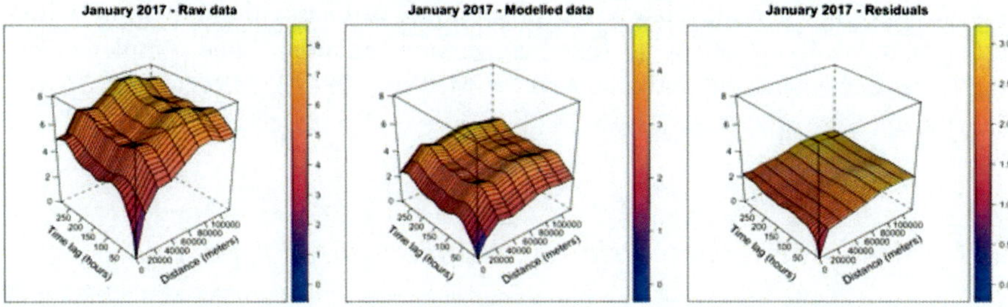

Fig. 7.7 Temporally extended semivariogram for January 2017. The same semivariogram of Fig. 7.6 computed up to 12 days

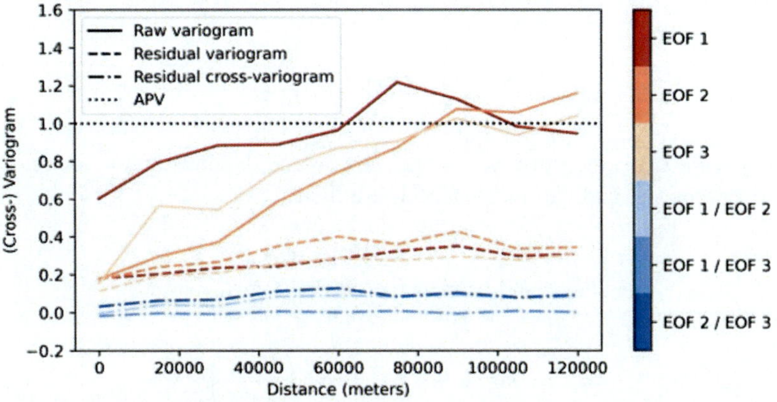

Fig. 7.8 Spatial variography of the first three components for MSWind 17. Spatial semivariograms of the raw data (solid line) and residuals (dashed line) after spatial modelling with ELM. Cross-semivariograms between residuals from two different components (dashed-dotted line). All components have a unitary sample variance (dotted line) due to the normalisation of the PC spatial coefficients

Table 7.3 Sample variances of the spatio-temporal semivariograms of Fig. 7.6. The sample variance is computed on the training data

s^2		Raw data	Modelled data	Residuals
June	2008	2.85	1.01	1.51
October	2013	6.66	2.34	3.46
April	2015	6.62	2.56	3.46
January	2017	6.61	3.40	2.71

7.3 Aeolian Energy Potential Estimation

The spatio-temporal wind field estimated in the previous section and its uncertainty are converted into wind turbine power potential. It is done with the help of a Taylor approximation of fitted wind speed conversion function. Some samples of the results are then visualised and analysed for MSWind 08–12, MSWind 13–16 and MSWind 17.

7.3.1 Wind Speed Conversion and Uncertainty Propagation

For convenience, let us note μ_Z the expectation and σ_Z^2 the variance of the wind speed $Z(\mathbf{s}_0, t_j)$, at a height $h_1 = 10$ [m] above the ground level. The wind speed $V(\mathbf{s}_0, t_j)$ at height h_2 is provided by the so-called *log-law*, given by

$$V(\mathbf{s}_0, t_j) = Z(\mathbf{s}_0, t_j) \cdot \frac{\ln \frac{h_2}{h_0}}{\ln \frac{h_1}{h_0}},$$

where $h_0 = h_0(\mathbf{s}_0)$ is the roughness of the terrain and depends on the location [13]. The expectation μ_V and variance σ_V^2 of V are then

$$\begin{aligned} \mu_V &= \mathbb{E}\left[V(\mathbf{s}_0, t_j)\right] = \mu_Z \cdot \frac{\ln \frac{h_2}{h_0}}{\ln \frac{h_1}{h_0}}, \\ \sigma_V^2 &= \mathrm{Var}\left[V(\mathbf{s}_0, t_j)\right] = \sigma_Z^2 \left(\frac{\ln \frac{h_2}{h_0}}{\ln \frac{h_1}{h_0}}\right)^2. \end{aligned} \tag{7.4}$$

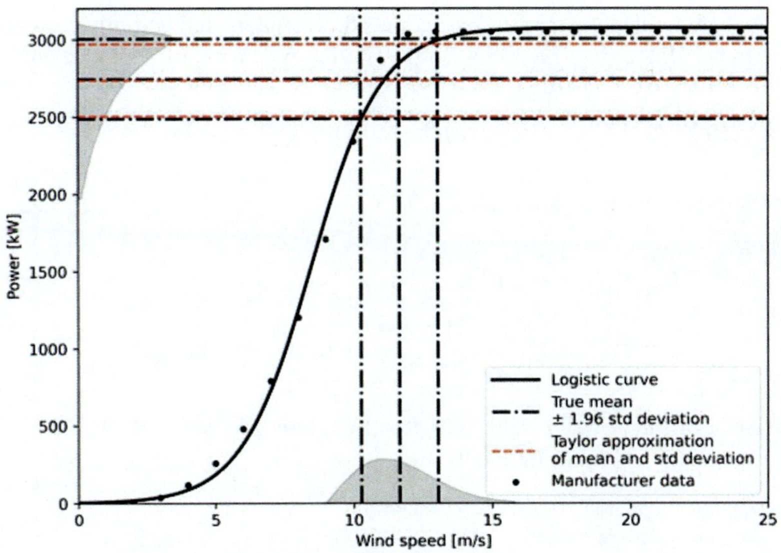

Fig. 7.9 Probabilistic approximation of power generation with a Taylor expansion. The power is considered as a logistic transformation of the wind speed, which is considered a random variable. The non-linear transformation completely modifies the distribution of wind speed. Data for Enercon E-101 wind turbine from the manufacturer are also reported

Once the wind speed has been estimated at the wind turbine height, it is converted to power. The logistic function has proven to be highly precise in fitting power curves, on simulated and manufacturers data [14, 15]. In addition to achieving a high degree of accuracy, the logistic function is easily derivable, which is convenient for approximation of uncertainty propagation. Suppose the power curve $P(v)$ of the turbine is given by a *three-parameters logistic function*

$$P(v) = \phi_1 S(v) \qquad \text{with} \qquad S(v) = \frac{1}{1 + \exp\left(\frac{\phi_2 - v}{\phi_3}\right)}.$$

An example of logistic curve is shown in Fig. 7.9. Note that the first and second derivatives of $P(v)$ are given by [16]

$$P'(v) = \frac{\phi_1}{\phi_3} S(v)(1 - S(v))$$
$$P''(v) = \frac{\phi_1}{\phi_3^2} S(v)(1 - S(v))(1 - 2\,S(v)).$$

The quantity $P(V)$ is a non-linear transformation of a random variable. Its expectation is not necessarily the non-linear transformed expectation of the wind speed $P(\mu_V)$; see Fig. 7.9. The whole distribution of V is necessary, although it is not available. Again, the delta method overcomes this problem and partially corrects this bias due to non-linearity. The second-order Taylor expansion of $P(V)$ around μ_V is

$$P(V) \simeq P(\mu_V) + P'(\mu_V)(V - \mu_V) + \frac{1}{2} P''(\mu_V)(V - \mu_V)^2.$$

Taking the expectation on both side,

$$\begin{aligned} \mathbb{E}[P(V)] &\simeq P(\mu_V) + \frac{1}{2} P''(\mu_V)\mathbb{E}\left[(V - \mu_V)^2\right] \\ &= \phi_1 S(\mu_V) \left[1 + \frac{1}{2\phi_3^2}(1 - S(\mu_V))(1 - 2\, S(\mu_V))\sigma_V^2\right]. \end{aligned} \tag{7.5}$$

The variance of the power is approximated by taking the variance of its first-order Taylor expansion, as higher moments are not available,

$$\mathrm{Var}\,[P(V)] \simeq (P'(\mu_V))^2 \sigma_V^2 = \frac{\phi_1^2}{\phi_3^2} S^2(\mu_V)(1 - S(\mu_V))^2 \sigma_V^2. \tag{7.6}$$

Let us remark that the logistic transformation completely modifies the variance—see Fig. 7.9. When wind speed is sufficiently high with a limited amount of variance to stay confidently in the plateau characterising the maximum of power between 15 and 25 [m/s], the power variance is small—in accordance with the approximation (7.6)—describing the high certainty in having the maximum of energy production. Similarly, when wind speed is sufficiently low with a limited amount of variance, the power is near zero with high confidence. Contrariwise, when the wind speed is in the transition phase of the logistic function, even with a reasonable amount of variance, the power is susceptible to fluctuate between its minimum and maximum with a high variance—mathematically described by a high derivative of $P(v)$ in the approximation (7.6).

Finally, the expected value and variance of the wind turbine power at each location $\mathbf{s}_0$ and each time t_j are estimated by replacing μ_Z and σ_Z^2 by $\hat{Z}$ and $\hat{\sigma}_P^2$ in Eqs. (7.4), and plug them in Eqs. (7.5) and (7.6).

7.3.2 Power Estimation for Switzerland

The aforementioned approximated conversion and uncertainty propagation are applied to the three modelled wind speed data sets. The power estimation is performed with an Enercon E-101 wind turbine at 100 [m] above the ground level [17]. The predicted wind speed data and its predicted variance are transformed, with height $h_2 = 100$ [m] and roughness h_0 derived from the Corine Land Cover maps of 2012 for the MSWind 08-12 and of 2018 for the remaining two data sets [18]. Transformed wind speed is removed when it is greater than 25 [m/s]—which is the

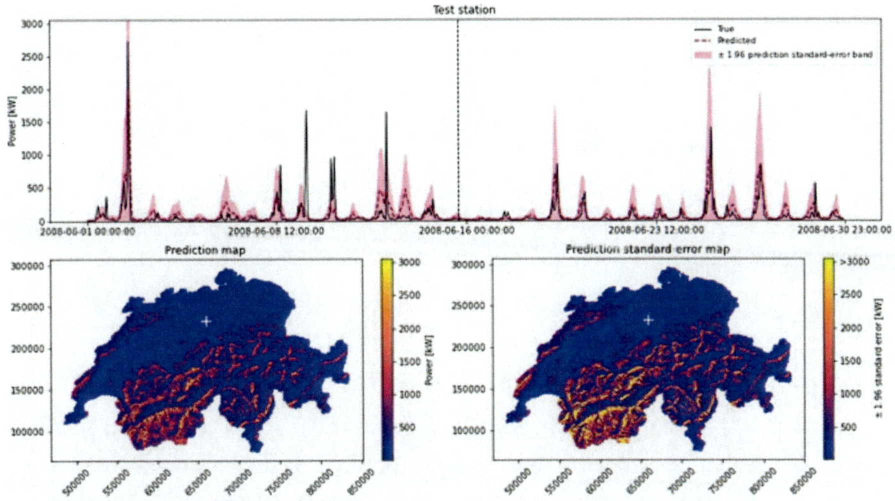

Fig. 7.10 Power prediction for the MSWind 08-12 data set: (top) The power time series at the MOA testing station obtained by passing the true wind speed in the logistic function (in black) and the predicted time series (in magenta) obtained using the Taylor expansion; (bottom left) The predicted map of power generation, at the fixed time indicated by the vertical dashed line in the temporal plot above, obtained as the expectation approximation (7.5); (bottom right) Map of the prediction standard error multiplied by 1.96 at the same time, obtained as the square root of the variance approximation (7.6)

cut-out wind speed of the turbine and then do not produce power. The manufacturer wind turbine power curve [17] is fitted with the R Package WindCurves [14], yielding $\phi_1 = 3075.31$, $\phi_2 = 8.47$ and $\phi_3 = 1.27$. Then, transformed wind speed and its variance are passed into Eqs. (7.5) and (7.6). One finally obtains an estimation of mean power accompanied by its variance on entire Switzerland for ten years.

Figures 7.10, 7.11 and 7.12 shows some samples of the results at the same locations and periods previously showed for wind speed modelling. Each Figure displays a partial power time series at a testing station, a prediction map at a fixed time and its corresponding UQ map. For comparison, the power obtained by passing the true wind speed measurement in the three-parameters logistic function $P(v)$ is added on the time series plots.

Globally, the leading behavioural changes of the true time series are captured, and it is contained in the ± 1.96 error bands for the three Figures. In Fig. 7.10 which displays the power time series for the MOA testing station, the model reproduces the daily cyclicity inherited from the wind speed behaviour during June 2008. On the maps, the spatial power patterns reproduce those of wind speed at the same time, likely because the wind is calm. Figure 7.11 provides more confusing time series at the CDF station. Remark that the true time series is within the magenta band most of the time, as expected. Interestingly, when production reaches its maximum potential defined by physical turbine characteristics, the error band sometimes shrinks, e.g.

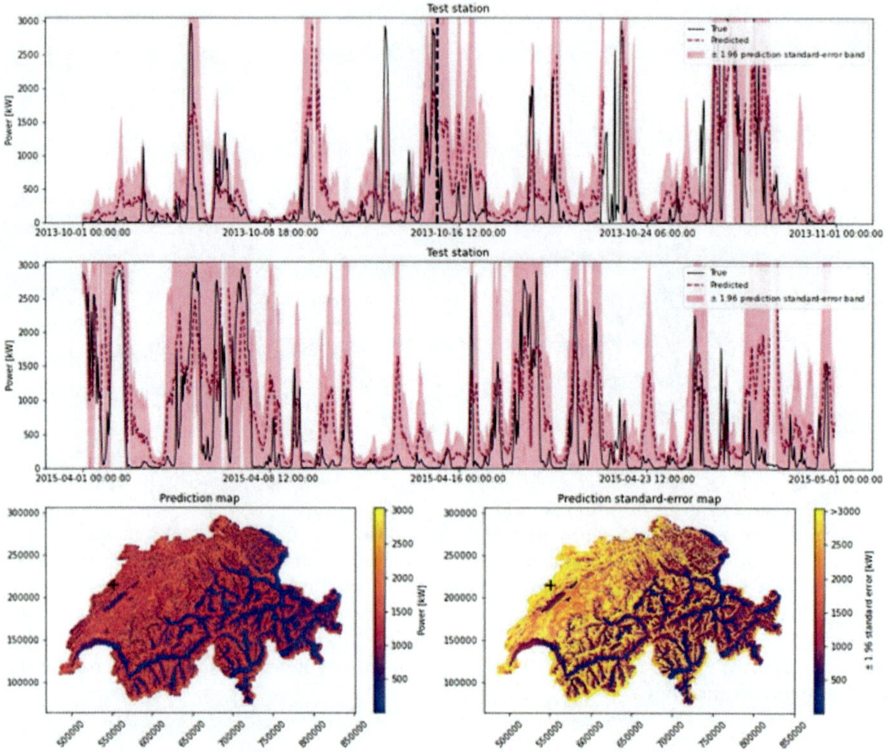

Fig. 7.11 Power prediction for the MSWind 13–16 data set: (top and centre) The power time series at the CDF testing station (in black) and the predicted time series (in magenta); (bottom left) The predicted map of power generation, at the fixed time indicated by the vertical dashed line in the temporal plot above; (bottom right) Map of the prediction standard error multiplied by 1.96 at the same time

the 2nd of April 2015. It was expected, due to the logistic transformation and its consequences on the variance behaviour stated previously. It also explains the more confusing aspect of this time series. The prediction map shows a good potential overall Switzerland, about 1500–2000 [kW]. However, these values are in the transition phase, which coupled with wind speed variance yield an uncertainty map with very high values. Therefore, the potential for Switzerland at this fixed time is not certain at all. Finally, Fig. 7.12 shows the BUS station in January 2017. The time series shows variability shrinking again. The maps provide a very interesting insight. An essential part of the Jura shows a very low uncertainty, while the power prediction is at its maximum. This behaviour is of particular interest for practical reasons as it allows the model to ensure high energy production. Some similar spots are also identifiable in the Western Plateau.

Further knowledge is extracted from these estimations and their UQ, and aeolian power potential is computed under several scenarios; see [19] for more details.

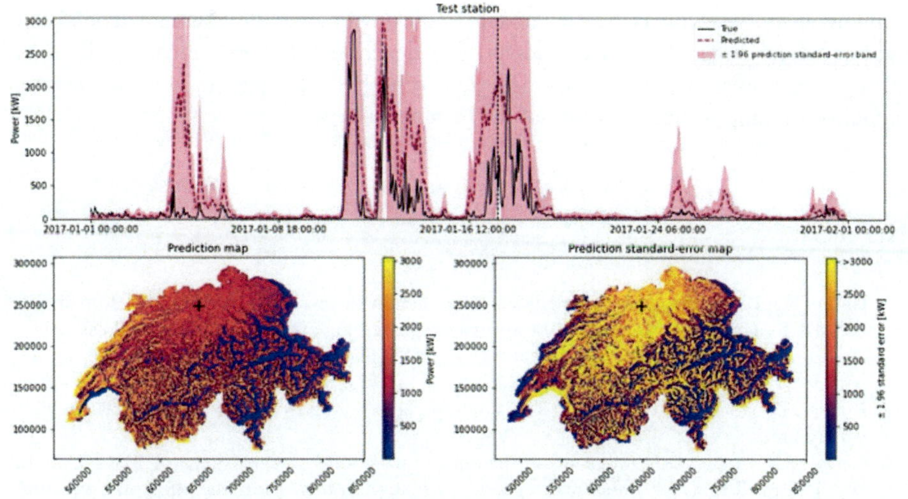

Fig. 7.12 Model prediction for the MSWind 17 data set: (top) The power time series at the BUS testing station (in black) and the predicted time series (in magenta); (bottom left) The predicted map of power generation, at the fixed time indicated by the vertical dashed line in the temporal plot above; (bottom right) Map of the prediction standard error multiplied by 1.96 at the same time

7.4 Summary

In this chapter, the single-output strategy of the spatio-temporal framework proposed in Chap. 5 is applied, adopting ELM to predict each spatial coefficient map individually. The variance estimates developed in Chap. 6 are used to extend UQ to the spatio-temporal framework. The prediction variance is estimated by a second model based on squared residuals that are log-transformed. Note that the ELM based variance estimate of this second model is also used to back-transform the results. These developments are practised on wind speed data. Finally, a comprehensive residuals analysis demonstrated that the model consistently predicts wind speed and quantifies its uncertainty, despite the complexity introduced by the hourly frequency of data and the relatively low number of training points.

The use of the regularised version of ELM provided an exciting visualisation tool—Fig. 7.1. Careful consideration of the regularisation parameter matrices contributes to providing insightful information about the spatio-temporal model to understand its behaviour, and furnish explainability of the models in terms of data interpretability. In this specific case, those insights are partly supported by previous EDA findings and confirmed by the residuals analysis.

Power generation is then approximated based on the modelled wind speed to assess renewable energy potential in Switzerland. As expected, the high variance propagated in the transition phase of the logistic function can lead to very uncertain predictions. An alternative way could be to do the spatio-temporal modelling directly on the power transformed data. However, a significant advantage of modelling the wind speed as a first step is the easy update of the power estimation, which depends on the turbine height and logistic parameters that describe turbine technical specificities through the power curve. For instance, this could help to generate multiple turbine scenarios or support decisions about the turbine selection.

References

1. Heskes T (1997) Practical confidence and prediction intervals. In: Mozer MC, Jordan MI, Petsche T (eds) Advances in neural information processing systems, vol 9. MIT Press, 1997, pp 176–182. http://papers.nips.cc/paper/1306-practical-confidence-and-prediction-intervals.pdf
2. Ruppert D, Wand MP, Carroll RJ (2003) Semiparametric regression, vol 12. Cambridge University Press
3. Carroll RJ, Ruppert D (1988) Transformation and weighting in regression, vol 30. CRC Press
4. Hall P, Carroll RJ (1989) Variance function estimation in regression: the effect of estimating the mean. J R Stat Soc: Ser B (Methodological) 51(1):3–14
5. Ver Hoef JM (2012) Who invented the delta method? Am Stat 66(2):124–127
6. Oehlert GW (1992) A note on the delta method. Am Stat 46(1):27–29
7. Golub GH, Heath M, Wahba G (1979) Generalized cross-validation as a method for choosing a good ridge parameter. Technometrics 21(2):215–223
8. Piegorsch WW (2015) Statistical data analytics: foundations for data mining, in- formatics, and knowledge discovery. Wiley
9. James G, Witten D, Hastie T, Tibshirani R (2013) An introduction to statistical learning, vol 112. Springer
10. Hastie T, Tibshirani R, Friedman J (2009) The elements of statistical learning: data mining, inference, and prediction, ser. Springer Series in Statistics. Springer. ISBN: 9780387848846
11. Kanevski M, Maignan M (2004) Analysis and modelling of spatial environmental data, vol 6501. EPFL Press
12. Chiles J-P, Delfiner P (2009) Geostatistics: modeling spatial uncertainty, vol 497. Wiley
13. Whiteman C (2000) Mountain meteorology: fundamentals and applications. Oxford University Press
14. Bokde N, Feijóo A, Villanueva D (2018) cd on the Weibull cumulative distribution function. Appl Sci 8(10):1757
15. Villanueva D, Feijóo AE (2016) Reformulation of parameters of the logistic function applied to power curves of wind turbines. Electric Power Syst Res 137:51–58
16. Minai AA, Williams RD (1993) On the derivatives of the sigmoid. Neural Netw 6(6):845–853
17. Wind-turbine-models.com, Enercon E-101 (2021). https://en.wind-turbine-models.com/turbines/130-enercon-e-101#datasheet. Accessed 30 Mar 2021
18. Grassi S, Veronesi F, Raubal M (2015) Satellite remote sensed data to improve the accuracy of statistical models for wind resource assessment. In: European wind energy association annual conference and exhibition 2015
19. Guignard F, Amato F, Walch A et al (2021) Spatio-temporal estimation of wind speed and wind power using machine learning: predictions, uncertainty and policy indications. Under review

Chapter 8
Conclusions, Perspectives and Recommendations

This concluding chapter summarises and underlines the main achievements of each research topic and presents some reflections on future research. It is organised as follows. Section 8.1 discusses Fisher-Shannon analysis treated in Chap. 4. The spatio-temporal interpolation framework presented in Chap. 5 and used in Chap. 7 is discussed in Sect. 8.2. Section 8.3 considers the ELM UQ developed in Chap. 6 and extended to the interpolation framework in Chap. 7. Finally, Sect. 8.4 discussed the application results obtained throughout the thesis.

8.1 Fisher-Shannon Analysis and Complexity Quantification

8.1.1 Thesis Achievements

The Chap. 4 discusses the Fisher-Shannon information method as an effective data exploration tool able to give diverse insights into complex non-stationary time series and spatio-temporal signals. The FSC was of particular interest. The Fisher-Shannon analysis was presented in a unified framework, and new interpretations of FSC were pointed out.

From a methodological point of view, FIM and FSC were computed in closed forms for several parametric distributions widely used in geo-environmental data analyses. Theoretical formulas for other random variables can be derived depending on the problem at hand. Analytical results derived for FSC presented quite elegant formulas. Mathematically speaking, FSC is a number that summaries a PDF for which location and scale parameters has no influence. On the one hand, the FSC was used as a statistical complexity measure. On the other hand, it was identified

F. Guignard, *On Spatio-Temporal Data Modelling and Uncertainty Quantification Using Machine Learning and Information Theory* , Springer Theses,
https://doi.org/10.1007/978-3-030-95231-0_8

as a SEP sensitivity measure and as a scale-independent non-Gaussianity measure, which both provide interpretation of this quantity.

Practically, it was attempted to investigate how this non-Gaussianity measure reacts to multimodality in a particular basic setting, yielding the unexpected results of being equal to the squared number of modes. It may provide a way to estimate the number of clusters in a given data set. Moreover, it was also shown on simulated data—by injecting noise into logistic map—how Fisher-Shannon analysis can be used to detect potential dynamic changes in a relatively robust manner—especially with FSC. The detection of potential Gaussian behaviour of FSC in the data was successfully shown in high-frequency wind speed data. The FSC was also used to discriminate wind profile behaviours in an urban zone by quantifying complexity at a given height. Other case studies were mentioned, particularly a climatic one, which may be a promising and valuable application.

Initially intended for tracking signal non-stationarity [6], the Fisher-Shannon method has been widely used in geosciences, as shown in Sect. 4.1. However, according to the thesis author opinion, its full potential is still unexploited and underestimated. Open source libraries written in R and Python to compute the three measures via a non-parametric KDE were provided to simplify the access to this method for environmental data analysis and foster reproducibility.

Finally, SEP, FIM and FSC are versatile information-based exploratory tools that can also be used in time series discrimination or, more generally, to generate time series features for clustering, modelling and forecasting.

8.1.2 Implications and Future Challenges

The Fisher-Shannon analysis is under-investigated. Many examples demonstrate its practical usefulness and let allow us to suspect a considerable potential for advanced EDA. It must be the attention of further investigations. In particular, in which sense can we think of non-Gaussianity about FSC? A part of the answer related to multimodality was just scratched with the thesis author investigations, while another part of the answer could lie in heavy-tailed distributions. Vignat and Bercher showed some results on the Student's t-distribution and the exponential power distribution, from which FSC is straight-forwardly obtained [6]. It could directly impact (environmental) risk assessment and, more generally, UQ.

In this thesis, the Fisher-Shannon analysis was used to extract knowledge from spatio-temporal data through temporal sliding windows, which, strictly speaking, is not a spatio-temporal tool. However, it can be easily adapted to spatial or spatio-temporal sliding windows, which was never investigated up to the author knowledge.

From a theoretical point of view, future studies should involve the generalisation of the Fisher-Shannon method to the multivariate case. It could also provide a manner to study temporal, spatial and spatio-temporal data, providing some stationarity of the underlying process. Moreover, several numerical investigations could be carried out for the KDE of the FIM. In particular, other estimates could be provided by resubstitution techniques as with entropy. Optimal bandwidth choice regarding asymptotic MSE of FIM—or even FSC—could be derived. Finally, theoretical and practical investigations on normal mixtures densities could open a way to study clustering problems and unsupervised learning.

8.2 Spatio-Temporal Interpolation with Machine Learning

8.2.1 Thesis Achievements

The Chap. 5 introduced a framework for spatio-temporal prediction, which is adaptable to any ML algorithm and allows the introduction of spatial covariates. Using the temporal formulation of EOFs to decompose the spatio-temporal signal into fixed temporal bases and stochastic spatial coefficients provides several key advantages. First, it allows to reconstruct spatio-temporal fields starting from spatially irregularly distributed measurements. Second, the framework can capture non-linear patterns in the data, as it models spatio-temporal fields as a linear combination of temporal bases with spatial coefficients maps, where the latter are obtained using a non-linear model. Third, non-stationarity, complex seasonal and other typical behaviours of high-frequency temporal data are captured in the temporal bases. Finally, note that this basis-function representation does not necessarily induce a separable model [1].

The framework was tested with a DL algorithm on simulated and real-world environmental data—temperature and wind speed—in a complicated spatial domain with complex terrain constraints. While the spatial prediction maps of the stochastic coefficients can be performed using any regression algorithm, DL algorithms are particularly well suited to solve this problem thanks to their automatic feature representation learning. Indeed, even if the traditional ML and geostatistical techniques could be used to model each single spatial coefficient map separately, the use of a single DL model allows the development of a network structure with multiple outputs to model them all coherently. Moreover, the recomposition of the spatio-temporal field can be executed through an additional layer embedded in the network, allowing to train the entire model to minimise a loss computed directly on the output signal.

Although DL seems to have numerous advantages over other regression algorithms in the context of the proposed spatio-temporal framework, the latter was also tested with ELM in Chap. 7, having in mind the aim of UQ assessment. However, the price to pay was to provide human-engineered spatio-temporal features based on heuristic rules to ELM to achieve comparable prediction performance, while DL does not require them. With both DL and ELM, it was shown that the proposed framework succeeds at recovering spatial, temporal and spatio-temporal dependencies.

8.2.2 Implications and Future Challenges

The proposed framework may be generalised to study other climate fields or environmental spatio-temporal phenomena—e.g. air pollution—or to solve missing data imputation problems in spatio-temporal datasets collected by satellites for earth observation or resulting from climate models.

As every ML model can be used with this approach, users interested in UQ can use methods that allow its explicit estimation. In this context, a solution was proposed and developed for ELM in this thesis, but similar development could be done for Gaussian Processes [4] or Random Forest [9–11], for instance. Nevertheless, careful developments have to be done to define the uncertainty propagation procedure through the framework's steps. Finally, additional fundamental studies should be conducted to extend this approach for spatio-temporal forecasting and multivariate analysis.

8.3 Uncertainty Quantification with Extreme Learning Machine

8.3.1 Thesis Achievements

The Chap. 6 discussed variance of (regularised) ELM under general hypothesis and its estimation through small ensembles of retrained ELMs under homoskedastic and heteroskedastic hypothesis. As ELM is nothing more than linear regression in a random feature space, analytical results can be derived by conditioning on the random input weights and biases.

Based on these formulas, several variance estimates independent of the noise distribution were provided for homoskedastic and heteroskedastic cases, for which a Python implementation was provided. Numerical simulations and empirical findings support formulas and estimate-related theoretical results.

The possibility of constructing accurate CI for $f(\mathbf{x}_0)$ and $\mathbb{E}[\hat{f}(\mathbf{x}_0)]$ despite the non-parametric, non-linear, and random nature of ELM was also shown in the Chap. 6. A detailed explanation of the bias/variance contribution in CI estimation was provided, and the fact that bias must be carefully considered to achieve satisfactory performances was highlighted. It is especially true in the regularised case, which introduces significant bias. In particular, bias was traded against variance, which can be estimated while preserving the predicting performance of the modelling, leading to credible uncertainty estimation. Also, as the variance estimates are distribution-free, it is reasonable to think that CI could be built with non-Gaussian noise distributional assumptions.

To summarise, the proposed ELM UQ results of Chap. 6 can be applied in any ELM modelling, which is a very general contribution to the ML literature. The randomness of ELM and the potential lack of understanding of the high dimensional input and projected spaces that partially motivated this estimation's development are quantified.

UQ with ELM was extended to the proposed spatio-temporal framework in Chap. 7. In this context, the limited number of spatial training points constituted an additional motivation for the UQ development. At the same time, it provided the foundations to a proper methodology to quantify the uncertainty of the space-time framework, potentially with other algorithms than ELM.

8.3.2 Implications and Future Challenges

Several aspects of ELM UQ still need to be investigated. From a theoretical perspective, the (asymptotical) normality of ELM ensembles should be proved or disproved. Additionally, random matrix theory—which already provided theoretical results for ELM [12]—can be investigated in the UQ context.

Sometimes it happens, that the residuals still contain spatial correlation [2, 3]. Therefore, future studies could also involve dependant data, e.g. by adapting heteroskedasticity and autocorrelation consistent (HAC) estimations of the full noise covariance matrix, in the temporal, spatial or spatio-temporal cases [5, 7, 13, 14]. It would allow a better quantification of uncertainty in such situations, where considering only the diagonal of the covariance matrix tends to underestimate it.

Finally, the estimation of heteroskedasticity/autocorrelation based on a first ELM model could be used to improve the prediction of a second ELM based on a weighted/generalised LS procedure. One could even imagine recursively reiterating the process until convergence by providing a variance estimation of this second ELM.

8.4 Application on Wind Phenomenon and Wind Energy Potential Estimation

8.4.1 Thesis Achievements

Two wind speed data case studies were carefully analysed: one data set from the MeteoSwiss network and another data set from the MoTUS experiment.

The quite unique MoTUS data are composed of seven one-hertz wind time series recorded at equidistant levels from 1.5 to 25.5 [m] above the ground in an urban area. Besides classical EDA tools, this data set was investigated with Fisher-Shannon analysis and wavelet variance. The results pointed out that wind structure is considerably modified in the urban canyon.

The FSC analysis based on daily moving window revealed a clustering tendency between the measurements below and above the building average height, suggesting different wind dynamics induced by the building layout. Clear correlations were also found between daily FSC and daily variance of sonic temperature at each level of the mast, which indicated that temperature variation could be an essential predictor for high-frequency wind speed complexity. Such correlation is larger for the lower anemometers, suggesting that ambient temperature is an important forcing of the wind speed variability in the vicinity of the ground. In addition, the FSC was calculated globally over each time series and showed a linear decrease with the height of the sensors. In particular, it was demonstrated that the relationship of wind speed with height is non-linear, as otherwise, the FSC would have been constant.

The wavelet variance analysis confirmed the clustering tendency. Two wavelet scale ranges identified with the microscale turbulence and synoptic mean flows were sensitive to two different behaviours. The lower sensors have higher wavelet variance at the microscale, while the higher ones have higher wavelet variance at the synoptic scale. It suggests that the wind speed variability in urban zones changes with height above the ground level for these two ranges.

These findings contribute to a better knowledge and understanding of the urban wind speed and its underlying mechanism governing the wind fluctuations depending on heights from the ground.

The MeteoSwiss data set covered Switzerland and was measured from 2008 to 2017 at a one-hour sampling period. This data set was explored with classical EDA but also with tools designed specifically for spatio-temporal data. It painted a rich picture of these data typical of the wind phenomenon, such as dependencies in time, space, and space-time, anisotropy, non-separability, but also complex high-frequency temporal behaviour such as non-stationarity, multicyclicity, and multiscale variability. Moreover, most of those characteristics dynamically change with time, and maybe space, space-time and observation scale.

The MeteoSwiss data were modelled with the proposed spatio-temporal framework to provide a prediction of wind speed and its standard error anywhere in Switzerland from 2008 to 2017. Moreover, the prediction variance was estimated by a second model. Subsequently, the results were converted into renewable power potential. A simple visual inspection of the converted results shows how the UQ can be as important as the prediction, if not more. It highlighted the importance of UQ to help the decision-making process in such applications.

8.4.2 Implications and Future Challenges

Modelling wind speed at 1 hour frequency among mountainous regions such as Switzerland is extremely hard. It would have been desirable to have results comparable with temperature data. However, data are very noisy and standard error bands rather large. It is not sure that other models could provide better results and should be a direction for future research. A promising candidate could be non-linear BHM, which could incorporate physical knowledge with equations of fluid mechanics.

The prediction variance was also of practical interest *per se*, and its understanding can provide insights into the data nature. In the case of wind speed, the variance estimations suggest that it depends on the three discussed regions of Switzerland, which is in good agreement with meteo-climatic observations. This understanding of variance depending on predictors could be improved by studying which covariate helps in its estimation with e.g. feature selection techniques.

The power approximation may be improved by supposing that the prediction distribution is not too far from a Gaussian, although it is unclear if this assumption is tenable. Indeed, the Taylor expansion could be obtained to higher orders, as the higher moments of the Gaussian distribution are known from its variance, and the higher derivatives of the logistic function are known and easily implementable [8].

Note also the young existence of the European Space Agency (ESA) satellite Aeolus, launched successfully on 22nd August 2018. Aeolus is the first wind LIDAR worldwide in space and directly observes wind profiles from the surface up to 30 [km] altitude on a global scale [15]. Data have been publicly available since mid-2020 and could be of great help in such wind speed studies and aeolian potential estimations.

References

1. Wikle, C., Zammit-Mangion, A., Cressie, N.: Spatio-temporal statistics with R, Ser. Chapman & Hall/CRC the R series. CRC Press, Taylor & Francis Group (2019)
2. Kanevski, M., Pozdnoukhov, A., Timonin, V.: Machine Learning for Spatial Environmental Data. EPFL Press (2009)
3. Kanevski, M., Maignan, M.: Analysis and modelling of spatial environmental data, vol 6501. EPFL Press (2004)
4. Rasmussen, C.E., Williams, C.K.: Gaussian Processes for Machine Learning. MIT Press (2006)
5. Amato, F., Guignard, F., Robert, S., Kanevski, M.: A novel framework for spatio-temporal prediction of environmental data using deep learning. Sci. Rep. **10**(1), 22–243 (2020). https://doi.org/10.1038/s41598-020-79148-7
6. Vignat, C., Bercher, J.-F.: Analysis of signals in the Fisher-Shannon information plane. Phys. Lett. A **312**(1), 27–33 (2003). ISSN: 0375-9601. https://doi.org/10.1016/S0375-9601(03)00570-X
7. Davidson R, MacKinnon JG et al (2004) Econometric theory and methods, vol 5. Oxford University Press, New York
8. Minai AA, Williams RD (1993) On the derivatives of the sigmoid. Neural Netw 6(6):845–853
9. Wager S, Hastie T, Efron B (2014) Confidence intervals for random forests: the jackknife and the infinitesimal jackknife. J Mach Learn Res 15(1):1625–1651
10. Mentch L, Hooker G (2016) Quantifying uncertainty in random forests via confidence intervals and hypothesis tests. J Mach Learn Res 17(1):841–881
11. Polimis K, Rokem A, Hazelton B (2017) Confidence intervals for random forests in python. J Open Sour Softw 2(19):124
12. Louart C, Liao Z, Couillet R et al (2018) A random matrix approach to neural networks. Ann Appl Probab 28(2):1190–1248

13. Newey WK, West KD (1986) A simple, positive semi-definite, heteroskedasticity and autocorrelationconsistent covariance matrix. Technical report, National Bureau of Economic Research
14. Kelejian HH, Prucha IR (2007) HAC estimation in a spatial framework. J Econom 140(1):131–154
15. ESA (European Space Agency) (2020) Aeolus mission. https://earth.esa.int/eogateway/missions/aeolus. Accessed 10 Dec 2020